AF797049

The Black Death and The Dancing Mania

Justus Friedrich Karl Hecker

Copyright Notice

The text in this book has been downloaded from the internet and has been extensively edited and typeset.

Copyright © 2006 Objective Systems Pty Ltd ACN 085 119 953

CONTENTS

I

Introduction

Justus Friedrich Karl Hecker was one of three generations of distinguished professors of medicine. His father, August Friedrich Hecker, a most industrious writer, first practised as a physician in Frankenhausen, and in 1790 was appointed Professor of Medicine at the University of Erfurt. In 1805 he was called to the like professorship at the University of Berlin. He died at Berlin in 1811.

Justus Friedrich Karl Hecker was born at Erfurt in January, 1795. He went, of course – being then ten years old – with his father to Berlin in 1805, studied at Berlin in the Gymnasium and University, but interrupted his studies at the age of eighteen to fight as a volunteer in the war for a renunciation of Napoleon and all his works. After Waterloo he went back to his studies, took his doctor's degree in 1817 with a treatise on the "Antiquities of Hydrocephalus," and became privat-docent in the Medical

Faculty of the Berlin University. His inclination was strong from the first towards the historical side of inquiries into Medicine. This caused him to undertake a "History of Medicine," of which the first volume appeared in 1822. It obtained rank for him at Berlin as Extraordinary Professor of the History of Medicine. This office was changed into an Ordinary professorship of the same study in 1834, and Hecker held that office until his death in 1850.

The office was created for a man who had a special genius for this form of study. It was delightful to himself, and he made it delightful to others. He is regarded as the founder of historical pathology. He studied disease in relation to the history of man, made his study yield to men outside his own profession an important chapter in the history of civilisation, and even took into account physical phenomena upon the surface of the globe as often affecting the movement and character of epidemics.

III

The account of "The Black Death" here translated by Dr. Babington was Hecker's first important work of this kind. It was published in 1832, and was followed in the same year by his account of "The Dancing Mania." The books here given are the two that first gave Hecker a wide reputation. Many other such treatises followed, among them, in 1865, a treatise on the "Great Epidemics of the Middle Ages." Besides his "History of Medicine," which, in its second volume, reached into the fourteenth century, and all his smaller treatises, Hecker wrote a large number of articles in Encyclopaedias and Medical Journals. Professor J.F.K. Hecker was, in a more interesting way, as busy as Professor A.F. Hecker, his father, had been. He transmitted the family energies to an only son, Karl von Hecker, born in 1827, who distinguished himself greatly as a Professor of Midwifery, and died in 1882.

Benjamin Guy Babington, the translator of these books of Hecker's, belonged also to a family in which the study of Medicine has passed from father to son, and both have been writers. B.G. Babington was the son of Dr. William Babington, who was physician to Guy's Hospital for some years before 1811, when the extent of his private practice caused him to retire. He died in 1833. His son, Benjamin Guy Babington, was educated at the Charterhouse, saw service as a midshipman, served for seven years in India, returned to England, graduated as physician at Cambridge in 1831. He distinguished himself by inquiries into the cholera epidemic in 1832, and translated these pieces of Hecker's in 1833, for publication by the Sydenham Society. He afterwards translated Hecker's other treatises on epidemics of the Middle Ages. Dr. B.G. Babington was Physician to Guy's Hospital from 1840 to 1855, and was a member of the Medical Council of the General Board of Health. He died on the 8th of April, 1866.

H.M.

THE BLACK DEATH

CHAPTER I

GENERAL OBSERVATIONS

***

That Omnipotence which has called the world with all its living creatures into one animated being, especially reveals Himself in the desolation of great pestilences. The powers of creation come into violent collision; the sultry dryness of the atmosphere; the subterraneous thunders; the mist of overflowing waters, are the harbingers of destruction. Nature is not satisfied with the ordinary alternations of life and death, and the destroying angel waves over man and beast his flaming sword.

These revolutions are performed in vast cycles, which the spirit of man, limited, as it is, to a narrow circle of perception, is unable to explore. They are, however, greater terrestrial events than any of those which proceed from the discord, the distress, or the passions of nations. By annihilations they awaken new life; and when the tumult above and below the earth is past, nature is renovated, and the mind awakens from torpor and depression to the consciousness of an intellectual existence.

Were it in any degree within the power of human research to draw up, in a vivid and connected form, an historical sketch of such mighty events, after the manner of the historians of wars and battles, and the migrations of nations, we might then arrive at clear views with respect to the mental development of the human race, and the ways of Providence would be more plainly discernible. It would then be demonstrable, that the mind of nations is deeply affected by the destructive conflict of the powers of nature, and that great disasters lead to striking changes in general civilisation. For all that exists in man, whether good or evil, is rendered conspicuous by the presence of great danger. His inmost feelings are roused – the thought of self-preservation masters his spirit – self-denial is put to severe proof, and wherever darkness and barbarism prevail, there the affrighted mortal flies to the idols of his superstition, and all laws, human and divine, are criminally violated.

In conformity with a general law of nature, such a state of excitement brings about a change, beneficial or detrimental, according to circumstances, so that nations either attain a higher degree of moral worth, or sink deeper in ignorance and vice. All this, however, takes place upon a much grander scale than through the ordinary vicissitudes of war and peace, or the rise and fall of empires, because the powers of nature themselves produce plagues, and subjugate the human will, which, in the contentions of nations, alone predominates.

CHAPTER II

THE DISEASE

***

The most memorable example of what has been advanced is afforded by a great pestilence of the fourteenth century, which desolated Asia, Europe, and Africa, and of which the people yet preserve the remembrance in gloomy traditions. It was an oriental plague, marked by inflammatory boils and tumours of the glands, such as break out in no other febrile disease. On account of these inflammatory boils, and from the black spots, indicatory of a putrid decomposition, which appeared upon the skin, it was called in Germany and in the northern kingdoms of Europe the Black Death, and in Italy, la mortalega grande, the Great Mortality.

Few testimonies are presented to us respecting its symptoms and its course, yet these are sufficient to throw light upon the form of the malady, and they are worthy of credence, from their coincidence with the signs of the same disease in modern times.

The imperial writer, Kantakusenos, whose own son, Andronikus, died of this plague in Constantinople, notices great imposthumes of the thighs and arms of those affected, which, when opened, afforded relief by the discharge of an offensive matter. Buboes, which are the infallible signs of the oriental plague, are thus plainly indicated, for he makes separate mention of smaller boils on the arms and in the face, as also in other parts of the body, and clearly distinguishes these from the blisters, which are no less produced by plague in all its forms. In many cases, black spots broke out all over the body, either single, or united and confluent.

These symptoms were not all found in every case. In many, one alone was sufficient to cause death, while some patients recovered, contrary to expectation, though afflicted with all. Symptoms of cephalic affection were frequent;

many patients became stupefied and fell into a deep sleep, losing also their speech from palsy of the tongue; others remained sleepless and without rest. The fauces and tongue were black, and as if suffused with blood; no beverage could assuage their burning thirst, so that their sufferings continued without alleviation until terminated by death, which many in their despair accelerated with their own hands. Contagion was evident, for attendants caught the disease of their relations and friends, and many houses in the capital were bereft even of their last inhabitant. Thus far the ordinary circumstances only of the oriental plague occurred. Still deeper sufferings, however, were connected with this pestilence, such as have not been felt at other times; the organs of respiration were seized with a putrid inflammation; a violent pain in the chest attacked the patient; blood was expectorated, and the breath diffused a pestiferous odour.

In the West, the following were the predominating symptoms on the eruption of this disease. An ardent fever, accompanied by an evacuation of blood, proved fatal in the first three days. It appears that buboes and inflammatory boils did not at first come out at all, but that the disease, in the form of carbuncular (anthrax-artigen) affection of the lungs, effected the destruction of life before the other symptoms were developed.

Thus did the plague rage in Avignon for six or eight weeks, and the pestilential breath of the sick, who expectorated blood, caused a terrible contagion far and near; for even the vicinity of those who had fallen ill of plague was certain death; so that parents abandoned their infected children, and all the ties of kindred were dissolved. After this period, buboes in the axilla and in the groin, and inflammatory boils all over the body, made their appearance; but it was not until seven months afterwards that some patients recovered with matured buboes, as in the ordinary milder form of plague.

Such is the report of the courageous Guy de Chauliac, who vindicated the honour of medicine, by bidding defiance to danger; boldly and constantly assisting the affected, and disdaining the excuse of his colleagues, who held the Arabian notion, that medical aid was unavailing, and that the contagion justified flight. He saw the plague twice in Avignon, first in the year 1348, from January to August, and then twelve years later, in the autumn, when it returned from Germany, and for nine months spread general distress and terror. The first time it raged chiefly among the poor, but in the year 1360, more among the higher classes. It now also destroyed a great many children, whom it had formerly spared, and but few women.

The like was seen in Egypt. Here also inflammation of the lungs was predominant, and destroyed quickly and infallibly, with burning heat and expectoration of blood. Here too the breath of the sick spread a deadly contagion, and human aid was as vain as it was destructive to those who approached the infected.

Boccacio, who was an eyewitness of its incredible fatality in Florence, the seat of the revival of science, gives a more lively description of the attack of the disease than his non-medical contemporaries.

It commenced here, not as in the East, with bleeding at the nose, a sure sign of inevitable death; but there took place at the beginning, both in men and women, tumours in the groin and in the axilla, varying in circumference up to the size of an apple or an egg, and called by the people, pest-boils (gavoccioli). Then there appeared similar tumours indiscriminately over all parts of the body, and black or blue spots came out on the arms or thighs, or on other parts, either single and large, or small and thickly studded. These spots proved equally fatal with the pest-boils, which had been from the first regarded as a sure sign of death. No power of medicine brought relief – almost all died within the first three days, some sooner, some later, after the appearance of these signs, and for the most part entirely without fever or other symptoms. The plague spread itself with the greater fury, as it communicated from the sick to the healthy, like fire among dry and oily fuel, and even contact with the clothes and other articles which had been used by the infected, seemed to induce the disease. As it advanced, not only men, but animals fell sick and shortly expired, if they had touched things belonging to the diseased or dead. Thus Boccacio himself saw two hogs on the rags of a person who had died of plague, after staggering about for a short time, fall down dead as if they had taken poison. In other places multitudes of dogs, cats, fowls, and other animals, fell victims to the contagion; and it is to be presumed that other epizootes among animals likewise took place, although the ignorant writers of the fourteenth century are silent on this point.

In Germany there was a repetition in every respect of the same phenomena. The infallible signs of the oriental bubo-plague with its inevitable contagion were found there as everywhere else; but the mortality was not nearly so great as in the other parts of Europe. The accounts do not all make mention of the spitting of blood, the diagnostic symptom of this fatal pestilence; we are not, however, thence to conclude that there was any considerable mitigation or modification of the disease, for we must not only take into account the defectiveness of the chronicles, but that isolated testimonies are often contradicted by many others. Thus the chronicles of Strasburg, which only take notice of boils and glandular swellings in the axillae and groins, are opposed by another account, according to which the mortal spitting of blood was met with in Germany; but this again is rendered suspicious, as the narrator postpones the death of those who were thus affected, to the sixth, and (even the) eighth day, whereas, no other author sanctions so long a course of the disease; and even in Strasburg, where a mitigation of the plague may, with most probability, be assumed since the year 1349, only 16,000 people were carried off, the generality expired by the third or fourth day. In Austria, and especially in Vienna, the plague was fully as malignant as anywhere, so that the patients who had red spots and black boils, as well as those afflicted with tumid glands, died about the third day; and lastly, very frequent sudden deaths occurred on the coasts of the North Sea and in Westphalia, without any further development of the malady.

To France, this plague came in a northern direction from Avignon, and was there more destructive than in Germany, so that in many places not more than two in twenty of the inhabitants survived. Many were struck, as if by lightning, and died on the spot, and this more frequently among the young and strong than the old; patients with enlarged glands in the axillae and groins scarcely survive two or three days; and no sooner did these fatal signs appear, than they bid adieu to the world, and sought consolation only in the absolution which Pope Clement VI. promised them in the hour of death.

In England the malady appeared, as at Avignon, with spitting of blood, and with the same fatality, so that the sick who were afflicted either with this symptom or with vomiting of blood, died in some cases immediately, in others within twelve hours, or at the latest two days. The inflammatory boils and buboes in the groins and axillae were recognised at once as prognosticating a fatal issue, and those were past all hope of recovery in whom they arose in numbers all over the body. It was not till towards the close of the plague that they ventured to open, by incision, these hard and dry boils, when matter flowed from them in small quantity, and thus, by compelling nature to a critical suppuration, many patients were saved. Every spot which the sick had touched, their breath, their clothes, spread the contagion; and, as in all other places, the attendants and friends who were either blind to their danger, or heroically despised it, fell a sacrifice to their sympathy. Even the eyes of the patient were considered a sources of contagion, which had the power of acting at a distance, whether on account of their unwonted lustre, or the distortion which they always suffer in plague, or whether in conformity with an ancient notion, according to which the sight was considered as the bearer of a demoniacal enchantment. Flight from infected cities seldom availed the fearful, for the germ of the disease adhered to them, and they fell sick, remote from assistance, in the solitude of their country houses.

Thus did the plague spread over England with unexampled rapidity, after it had first broken out in the county of Dorset, whence it advanced through the counties of Devon and Somerset, to Bristol, and thence reached Gloucester, Oxford and London. Probably few places escaped, perhaps not any; for the annuals of contemporaries report that throughout the land only a tenth part of the inhabitants remained alive.

From England the contagion was carried by a ship to Bergen, the capital of Norway, where the plague then broke out in its most frightful form, with vomiting of blood; and throughout the whole country, spared not more than a third of the inhabitants. The sailors found no refuge in their ships; and vessels were often seen driving about on the ocean and drifting on shore, whose crews had perished to the last man.

In Poland the affected were attacked with spitting blood, and died in a few days in such vast numbers, that, as it has been affirmed, scarcely a fourth of the inhabitants were left.

Finally, in Russia the plague appeared two years later than in Southern Europe; yet here again, with the same symptoms as elsewhere. Russian contemporaries have recorded that it began with rigor, heat, and darting pain in the shoulders and back; that it was accompanied by spitting of blood, and terminated fatally in two, or at most three days. It is not till the year 1360 that we find buboes mentioned as occurring in the neck, in the axillae, and in the groins, which are stated to have broken out when the spitting of blood had continued some time. According to the experience of Western Europe, however, it cannot be assumed that these symptoms did not appear at an earlier period.

Thus much, from authentic sources, on the nature of the Black Death. The descriptions which have been communicated contain, with a few unimportant exceptions, all the symptoms of the oriental plague which have been observed in more modern times. No doubt can obtain on this point. The facts are placed clearly before our eyes. We must, however, bear in mind that this violent disease does not always appear in the same form, and that while the essence of the poison which it produces, and which is separated so abundantly from the body of the patient, remains unchanged, it is proteiform in its varieties, from the almost imperceptible vesicle, unaccompanied by fever, which exists for some time before it extends its poison inwardly, and then excites fever and buboes, to the fatal form in which carbuncular inflammations fall upon the most important viscera.

Such was the form which the plague assumed in the fourteenth century, for the accompanying chest affection which appeared in all the countries whereof we have received any account, cannot, on a comparison with similar and familiar symptoms, be considered as any other than the inflammation of the lungs of modern medicine, a disease which at present only appears sporadically, and, owing to a putrid decomposition of the fluids, is probably combined with hemorrhages from the vessels of the lungs. Now, as every carbuncle, whether it be cutaneous or internal, generates in abundance the matter of contagion which has given rise to it, so, therefore, must the breath of the affected have been poisonous in this plague, and on this account its power of contagion wonderfully increased; wherefore the opinion appears incontrovertible, that owing to the accumulated numbers of the diseased, not only individual chambers and houses, but whole cities were infected, which, moreover, in the Middle Ages, were, with few exceptions, narrowly built, kept in a filthy state, and surrounded with stagnant ditches. Flight was, in consequence, of no avail to the timid; for even though they had sedulously avoided all communication with the diseased and the suspected, yet their clothes were saturated with the pestiferous atmosphere, and every inspiration imparted to them the seeds of the destructive malady, which, in the greater number of cases, germinated with but too much fertility. Add to which, the usual propagation of the plague through clothes, beds, and a thousand other things to which the pestilential poison adheres – a propagation which, from want of caution, must have been infinitely multiplied; and since articles of this kind, removed from the access of air, not only retain the matter of contagion for an indefinite period, but also increase its activity and engender it like a living being, frightful ill-consequences followed for many years after the first fury of the pestilence was past.

The affection of the stomach, often mentioned in vague terms, and occasionally as a vomiting of blood, was doubtless only a subordinate symptom, even if it be admitted that actual hematemesis did occur. For the difficulty of distinguishing a flow of blood from the stomach, from a pulmonic expectoration of that fluid, is, to non-medical men, even in common cases, not inconsiderable. How much greater then must it have been in so terrible a disease, where assistants could not venture to approach the sick without exposing themselves to certain death? Only two medical descriptions of the malady have reached us, the one by the brave Guy de Chauliac, the other by

Raymond Chalin de Vinario, a very experienced scholar, who was well versed in the learning of the time. The former takes notice only of fatal coughing of blood; the latter, besides this, notices epistaxis, hematuria, and fluxes of blood from the bowels, as symptoms of such decided and speedy mortality, that those patients in whom they were observed usually died on the same or the following day.

That a vomiting of blood may not, here and there, have taken place, perhaps have been even prevalent in many places, is, from a consideration of the nature of the disease, by no means to be denied; for every putrid decomposition of the fluids begets a tendency to hemorrhages of all kinds. Here, however, it is a question of historical certainty, which, after these doubts, is by no means established. Had not so speedy a death followed the expectoration of blood, we should certainly have received more detailed intelligence respecting other hemorrhages; but the malady had no time to extend its effects further over the extremities of the vessels. After its first fury, however, was spent, the pestilence passed into the usual febrile form of the oriental plague. Internal carbuncular inflammations no longer took place, and hemorrhages became phenomena, no more essential in this than they are in any other febrile disorders. Chalin, who observed not only the great mortality of 1348, and the plague of 1360, but also that of 1373 and 1382, speaks moreover of affections of the throat, and describes the back spots of plague patients more satisfactorily than any of his contemporaries. The former appeared but in few cases, and consisted in carbuncular inflammation of the gullet, with a difficulty of swallowing, even to suffocation, to which, in some instances, was added inflammation of the ceruminous glands of the ears, with tumours, producing great deformity. Such patients, as well as others, were affected with expectoration of blood; but they did not usually die before the sixth, and, sometimes, even as late as the fourteenth day. The same occurrence, it is well known, is not uncommon in other pestilences; as also blisters on the surface of the body, in different places, in the vicinity of which, tumid glands and inflammatory boils, surrounded by discoloured and black streaks, arose, and thus indicated the reception of the poison. These streaked spots were called, by an apt comparison, the girdle, and this appearance was justly considered extremely dangerous.

CHAPTER III

CAUSES – SPREAD

***

An inquiry into the causes of the Black Death will not be without important results in the study of the plagues which have visited the world, although it cannot advance beyond generalisation without entering upon a field hitherto uncultivated, and, to this hour entirely unknown. Mighty revolutions in the organism of the earth, of which we have credible information, had preceded it. From China to the Atlantic, the foundations of the earth were shaken – throughout Asia and Europe the atmosphere was in commotion, and endangered, by its baneful influence, both vegetable and animal life.

The series of these great events began in the year 1333, fifteen years before the plague broke out in Europe: they first appeared in China. Here a parching drought, accompanied by famine, commenced in the tract of country watered by the rivers Kiang and Hoai. This was followed by such violent torrents of rain, in and about Kingsai, at that time the capital of the empire, that, according to tradition, more than 400,000 people perished in the floods. Finally the mountain Tsincheou fell in, and vast clefts were formed in the earth. In the succeeding year (1334), passing over fabulous traditions, the neighbourhood of Canton was visited by inundations; whilst in Tche, after an unexampled drought, a plague arose, which is said to have carried off about 5,000,000 of people. A few months afterwards an earthquake followed, at and near Kingsai; and subsequent to the falling in of the mountains of Ki-ming-chan, a lake was formed of more than a hundred leagues in circumference, where, again, thousands found their grave. In Houkouang and Honan, a drought prevailed for five months; and innumerable swarms of locusts destroyed the vegetation; while famine and pestilence, as usual, followed in their train. Connected accounts of the condition of Europe before this great catastrophe are not to be expected from the writers of the fourteenth century. It is remarkable, however, that simultaneously with a drought and renewed floods in China, in 1336, many uncommon atmospheric phenomena, and in the winter, frequent thunderstorms, were observed in the north of France; and so early as the eventful year of 1333 an eruption of Etna took place. According to the Chinese annuals, about 4,000,000 of people perished by famine in the neighbourhood of Kiang in 1337; and deluges, swarms of locusts, and an earthquake which lasted six days, caused incredible devastation. In the same year, the first swarms of locusts appeared in Franconia, which were succeeded in the following year by myriads of these insects. In 1338 Kingsai was visited by an earthquake of ten days' duration; at the same time France suffered from a failure in the harvest; and thenceforth, till the year 1342, there was in China a constant succession of inundations, earthquakes, and famines. In the same year great floods occurred in the vicinity of the Rhine and in France, which could not be attributed to rain alone; for, everywhere, even on tops of mountains, springs were seen to burst forth, and dry tracts were laid under water in an inexplicable manner. In the following year, the mountain Hong-tchang, in China, fell in, and caused a destructive deluge; and in Pien-tcheon and Leang-tcheou, after three months' rain, there followed unheard-of inundations, which destroyed seven cities. In Egypt and Syria, violent earthquakes took place; and in China they became, from this time, more and more frequent; for they recurred, in 1344, in Ven-tcheou, where the sea overflowed in consequence; in 1345, in Ki-tcheou, and in both the following years in Canton, with subterraneous thunder. Meanwhile, floods and famine devastated various districts, until 1347, when the fury of the elements subsided in China.

The signs of terrestrial commotions commenced in Europe in the year 1348, after the intervening districts of country in Asia had probably been visited in the same manner.

On the island of Cyprus, the plague from the East had already broken out; when an earthquake shook the foundations of the island, and was accompanied by so frightful a hurricane, that the inhabitants who had slain their Mahometan slaves, in order that they might not themselves be subjugated by them, fled in dismay, in all directions. The sea overflowed – the ships were dashed to pieces on the rocks, and few outlived the terrific event, whereby this fertile and blooming island was converted into a desert. Before the earthquake, a pestiferous wind spread so poisonous an odour, that many, being overpowered by it, fell down suddenly and expired in dreadful agonies.

This phenomenon is one of the rarest that has ever been observed, for nothing is more constant than the composition of the air; and in no respect has nature been more careful in the preservation of organic life. Never have naturalists discovered in the atmosphere foreign elements, which, evident to the senses, and borne by the winds, spread from land to land, carrying disease over whole portions of the earth, as is recounted to have taken place in the year 1348. It is, therefore, the more to be regretted, that in this extraordinary period, which, owing to the low condition of science, was very deficient in accurate observers, so little that can be depended on respecting those uncommon occurrences in the air, should have been recorded. Yet, German accounts say expressly, that a thick, stinking mist advanced from the East, and spread itself over Italy; and there could be no deception in so palpable a phenomenon. The credibility of unadorned traditions, however little they may satisfy physical research, can scarcely be called in question when we consider the connection of events; for just at this time earthquakes were more general than they had been within the range of history. In thousands of places chasms were formed, from whence arose noxious vapours; and as at that time natural occurrences were transformed into miracles, it was reported, that a fiery meteor, which descended on the earth far in the East, had destroyed everything within a circumference of more than a hundred leagues, infecting the air far and wide. The consequences of innumerable floods contributed to the same effect; vast river districts had been converted into swamps; foul vapours arose everywhere, increased by the odour of putrified locusts, which had never perhaps darkened the sun in thicker swarms, and of countless corpses, which even in the well-regulated countries of Europe, they knew not how to remove quickly enough out of the sight of the living. It is probable, therefore, that the atmosphere contained foreign, and sensibly perceptible, admixtures to a great extent, which, at least in the lower regions, could not be decomposed, or rendered ineffective by separation.

Now, if we go back to the symptoms of the disease, the ardent inflammation of the lungs points out, that the organs of respiration yielded to the attack of an atmospheric poison – a poison which, if we admit the independent origin of the Black Plague at any one place of the globe, which, under such extraordinary circumstances, it would be difficult to doubt, attacked the course of the circulation in as hostile a manner as that which produces inflammation of the spleen, and other animal contagions that cause swelling and inflammation of the lymphatic glands.

Pursuing the course of these grand revolutions further, we find notice of an unexampled earthquake, which, on the 25th January, 1348, shook Greece, Italy, and the neighbouring countries. Naples, Rome, Pisa, Bologna, Padua, Venice, and many other cities, suffered considerably; whole villages were swallowed up. Castles, houses, and churches were overthrown, and hundreds of people were buried beneath their ruins. In Carinthia, thirty villages, together with all the churches, were demolished; more than a thousand corpses were drawn out of the rubbish;

the city of Villach was so completely destroyed that very few of its inhabitants were saved; and when the earth ceased to tremble it was found that mountains had been moved from their positions, and that many hamlets were left in ruins. It is recorded that during this earthquake the wine in the casks became turbid, a statement which may be considered as furnishing proof that changes causing a decomposition of the atmosphere had taken place; but if we had no other information from which the excitement of conflicting powers of nature during these commotions might be inferred, yet scientific observations in modern times have shown that the relation of the atmosphere to the earth is changed by volcanic influences. Why then, may we not, from this fact, draw retrospective inferences respecting those extraordinary phenomena?

Independently of this, however, we know that during this earthquake, the duration of which is stated by some to have been a week, and by others a fortnight, people experienced an unusual stupor and headache, and that many fainted away.

These destructive earthquakes extended as far as the neighbourhood of Basle, and recurred until the year 1360 throughout Germany, France, Silesia, Poland, England, and Denmark, and much further north.

Great and extraordinary meteors appeared in many places, and were regarded with superstitious horror. A pillar of fire, which on the 20th of December, 1348, remained for an hour at sunrise over the pope's palace in Avignon; a fireball, which in August of the same year was seen at sunset over Paris, and was distinguished from similar phenomena by its longer duration, not to mention other instances mixed up with wonderful prophecies and omens, are recorded in the chronicles of that age.

The order of the seasons seemed to be inverted; rains, flood, and failures in crops were so general that few places were exempt from them; and though an historian of this century assure us that there was an abundance in the granaries and storehouses, all his contemporaries, with one voice, contradict him. The consequences of failure in the crops were soon felt, especially in Italy and the surrounding countries, where, in this year, a rain, which continued for four months, had destroyed the seed. In the larger cities they were compelled, in the spring of 1347, to have recourse to a distribution of bread among the poor, particularly at Florence, where they erected large bakehouses, from which, in April, ninety-four thousand loaves of bread, each of twelve ounces in weight, were daily dispensed. It is plain, however, that humanity could only partially mitigate the general distress, not altogether obviate it.

Diseases, the invariable consequence of famine, broke out in the country as well as in cities; children died of hunger in their mother's arms – want, misery, and despair were general throughout Christendom.

Such are the events which took place before the eruption of the Black Plague in Europe. Contemporaries have explained them after their own manner, and have thus, like their posterity, under similar circumstances, given a proof that mortals possess neither senses nor intellectual powers sufficiently acute to comprehend the phenomena produced by the earth's organism, much less scientifically to understand their effects. Superstition, selfishness in a thousand forms, the presumption of the schools, laid hold of unconnected facts. They vainly thought to comprehend the whole in the individual, and perceived not the universal spirit which, in intimate union with the mighty powers of nature, animates the movements of all existence, and permits not any phenomenon to originate from isolated causes. To attempt, five centuries after that age of desolation, to point out the causes of a cosmical commotion, which has never recurred to an equal extent, to indicate scientifically the influences, which called forth so terrific a poison in the bodies of men and animals, exceeds the limits of human understanding. If we are even now unable, with all the varied resources of an extended knowledge of nature, to define that condition of the atmosphere by which pestilences are generated, still less can we pretend to reason retrospectively from the nineteenth to the fourteenth century; but if we take a general view of the occurrences, that century will give us copious information, and, as applicable to all succeeding times, of high importance.

In the progress of connected natural phenomena from east to west, that great law of nature is plainly revealed which has so often and evidently manifested itself in the earth's organism, as well as in the state of nations dependent upon it. In the inmost depths of the globe that impulse was given in the year 1333, which in uninterrupted succession for six and twenty years shook the surface of the earth, even to the western shores of Europe. From the very beginning the air partook of the terrestrial concussion, atmospherical waters overflowed the land, or its plants and animals perished under the scorching heat. The insect tribe was wonderfully called into life, as if animated beings were destined to complete the destruction which astral and telluric powers had begun. Thus did this dreadful work of nature advance from year to year; it was a progressive infection of the zones, which exerted a powerful influence both above and beneath the surface of the earth; and after having been perceptible in slighter indications, at the commencement of the terrestrial commotions in China, convulsed the whole earth.

The nature of the first plague in China is unknown. We have no certain intelligence of the disease until it entered the western countries of Asia. Here it showed itself as the Oriental plague, with inflammation of the lungs; in which form it probably also may have begun in China, that is to say, as a malady which spreads, more than any other, by contagion – a contagion that, in ordinary pestilences, requires immediate contact, and only under favourable circumstances of rare occurrence is communicated by the mere approach to the sick. The share which this cause had in the spreading of the plague over the whole earth was certainly very great; and the opinion that the Black Death might have been excluded from Western Europe by good regulations, similar to those which are now in use, would have all the support of modern experience, provided it could be proved that this plague had been actually imported from the East, or that the Oriental plague in general, whenever it appears in Europe, has its origin in Asia or Egypt. Such a proof, however, can by no means be produced so as to enforce conviction; for it would involve the impossible assumption, either that there is no essential difference between the degree of civilisation of the European nations, in the most ancient and in modern times, or that detrimental circumstances, which have yielded only to the civilisation of human society and the regular cultivation of countries, could not formerly keep up the glandular plague.

The plague was, however, known in Europe before nations were united by the bonds of commerce and social intercourse; hence there is ground for supposing that it sprang up spontaneously, in consequence of the rude manner of living and the uncultivated state of the earth, influences which peculiarly favour the origin of severe diseases. Now we need not go back to the earlier centuries, for the fourteenth itself, before it had half expired, was visited by five or six pestilences.

If, therefore, we consider the peculiar property of the plague, that in countries which it has once visited it remains for a long time in a milder form, and that the epidemic influences of 1342, when it had appeared for the last time, were particularly favourable to its unperceived continuance, till 1348, we come to the notion that in this eventful year also the germs of plague existed in Southern Europe, which might be vivified by atmospherical deteriorations; and that thus, at least in part, the Black Plague may have originated in Europe itself. The corruption of the atmosphere came from the East; but the disease itself came not upon the wings of the wind, but was only excited and increased by the atmosphere where it had previously existed.

This source of the Black Plague was not, however, the only one; for far more powerful than the excitement of the latent elements of the plague by atmospheric influences was the effect of the contagion communicated from one people to another on the great roads and in the harbours of the Mediterranean. From China the route of the caravans lay to the north of the Caspian Sea, through Central Asia, to Tauris. Here ships were ready to take the produce of the East to Constantinople, the capital of commerce, and the medium of connection between Asia, Europe, and Africa. Other caravans went from India to Asia Minor, and touched at the cities south of the Caspian Sea, and, lastly, from Bagdad through Arabia to Egypt; also the maritime communication on the Red Sea, from India to Arabia and Egypt, was not inconsiderable. In all these directions contagion made its way; and, doubtless, Constantinople and the harbours of Asia Minor are to be regarded as the foci of infection, whence it radiated to the most distant seaports and islands.

To Constantinople the plague had been brought from the northern coast of the Black Sea, after it had depopulated the countries between those routes of commerce, and appeared as early as 1347 in Cyprus, Sicily, Marseilles, and some of the seaports of Italy. The remaining islands of the Mediterranean, particularly Sardinia, Corsica, and Majorca, were visited in succession. Foci of contagion existed also in full activity along the whole southern coast of Europe; when, in January, 1348, the plague appeared in Avignon, and in other cities in the south of

France and north of Italy, as well as in Spain.

The precise days of its eruption in the individual towns are no longer to be ascertained; but it was not simultaneous; for in Florence the disease appeared in the beginning of April, in Cesena the 1st June, and place after place was attacked throughout the whole year; so that the plague, after it had passed through the whole of France and Germany - where, however, it did not make its ravages until the following year - did not break out till August in England, where it advanced so gradually, that a period of three months elapsed before it reached London. The northern kingdoms were attacked by it in 1349; Sweden, indeed, not until November of that year, almost two years after its eruption in Avignon. Poland received the plague in 1349, probably from Germany, if not from the northern countries; but in Russia it did not make its appearance until 1351, more than three years after it had broken out in Constantinople. Instead of advancing in a northwesterly direction from Tauris and from the Caspian Sea, it had thus made the great circuit of the Black Sea, by way of Constantinople, Southern and Central Europe, England, the northern kingdoms, and Poland, before it reached the Russian territories, a phenomenon which has not again occurred with respect to more recent pestilences originating in Asia.

Whether any difference existed between the indigenous plague, excited by the influence of the atmosphere, and that which was imported by contagion, can no longer be ascertained from facts; for the contemporaries, who in general were not competent to make accurate researches of this kind, have left no data on the subject. A milder and a more malignant form certainly existed, and the former was not always derived from the latter, as is to be supposed from this circumstance – that the spitting of blood, the infallible diagnostic of the latter, on the first breaking out of the plague, is not similarly mentioned in all the reports; and it is therefore probable that the milder form belonged to the native plague – the more malignant, to that introduced by contagion. Contagion was, however, in itself, only one of many causes which gave rise to the Black Plague.

This disease was a consequence of violent commotions in the earth's organism – if any disease of cosmical origin can be so considered. One spring set a thousand others in motion for the annihilation of living beings, transient or permanent, of mediate or immediate effect. The most powerful of all was contagion; for in the most distant countries, which had scarcely yet heard the echo of the first concussion, the people fell a sacrifice to organic poison – the untimely offspring of vital energies thrown into violent commotion.

CHAPTER IV

MORTALITY

***

We have no certain measure by which to estimate the ravages of the Black Plague, if numerical statements were wanted, as in modern times. Let us go back for a moment to the fourteenth century. The people were yet but little civilised. The Church had indeed subdued them; but they all suffered from the ill consequences of their original rudeness. The dominion of the law was not yet confirmed. Sovereigns had everywhere to combat powerful enemies to internal tranquillity and security. The cities were fortresses for their own defence. Marauders encamped on the roads. The husbandman was a feudal slave, without possessions of his own. Rudeness was general, humanity as yet unknown to the people. Witches and heretics were burned alive. Gentle rulers were contemned as weak; wild passions, severity and cruelty, everywhere predominated. Human life was little regarded. Governments concerned not themselves about the numbers of their subjects, for whose welfare it was incumbent on them to provide. Thus, the first requisite for estimating the loss of human life, namely, a knowledge of the amount of the population, is altogether wanting; and, moreover, the traditional statements of the amount of this loss are so vague, that from this source likewise there is only room for probable conjecture.

Cairo lost daily, when the plague was raging with its greatest violence, from 10,000 to 15,000; being as many as, in modern times, great plagues have carried off during their whole course. In China, more than thirteen millions are said to have died; and this is in correspondence with the certainly exaggerated accounts from the rest of Asia. India was depopulated. Tartary, the Tartar kingdom of Kaptschak, Mesopotamia, Syria, Armenia, were covered with dead bodies – the Kurds fled in vain to the mountains. In Caramania and Caesarea none were left alive. On the roads – in the camps – in the caravansaries – unburied bodies alone were seen; and a few cities only (Arabian historians name Maarael-Nooman, Schisur, and Harem) remained, in an unaccountable manner, free. In Aleppo, 500 died daily; 22,000 people, and most of the animals, were carried off in Gaza, within six weeks. Cyprus lost almost all its inhabitants; and ships without crews were often seen in the Mediterranean, as afterwards in the North Sea, driving about, and spreading the plague wherever they went on shore. It was reported to Pope Clement, at Avignon, that throughout the East, probably with the exception of China, 23,840,000 people had fallen victims to the plague. Considering the occurrences of the fourteenth and fifteenth centuries, we might, on first view, suspect the accuracy of this statement. How (it might be asked) could such great wars have been carried on – such powerful efforts have been made; how could the Greek Empire, only a hundred years later, have been overthrown, if the people really had been so utterly destroyed?

This account is nevertheless rendered credible by the ascertained fact, that the palaces of princes are less accessible to contagious diseases than the dwellings of the multitude; and that in places of importance, the influx from those districts which have suffered least, soon repairs even the heaviest losses. We must remember, also, that we do not gather much from mere numbers without an intimate knowledge of the state of society. We will therefore confine ourselves to exhibiting some of the more credible accounts relative to European cities.

To which may be added–

This short catalogue might, by a laborious and uncertain calculation, deduced from other sources, be easily further multiplied, but would still fail to give a true picture of the depopulation which took place. Lubeck, at that time the Venice of the North, which could no longer contain the multitudes that flocked to it, was thrown into such consternation on the eruption of the plague, that the citizens destroyed themselves as if in frenzy.

In Florence there died of the Black Plague	60,000
In Venice	100,000
In Marseilles, in one month	16,000
In Siena	70,000
In Paris	50,000
In St. Denys	14,000
In Avignon	60,000
In Strasburg	16,000
In Lubeck	9,000
In Basle	14,000
In Erfurt, at least	16,000
In Weimar	5,000
In Limburg	2,500
In London, at least	100,000
In Norwich	51,100

Franciscan Friars in German	124,434
Minorites in Italy	30,000

Merchants whose earnings and possessions were unbounded, coldly and willingly renounced their earthly goods. They carried their treasures to monasteries and churches, and laid them at the foot of the altar; but gold had no charms for the monks, for it brought them death. They shut their gates; yet, still it was cast to them over the convent walls. People would brook no impediment to the last pious work to which they were driven by despair. When the plague ceased, men thought they were still wandering among the dead, so appalling was the livid aspect of the survivors, in consequence of the anxiety they had undergone, and the unavoidable infection of the air. Many other cities probably presented a similar appearance; and it is ascertained that a great number of small country towns and villages, which have been estimated, and not too highly, at 200,000, were bereft of all their inhabitants.

In many places in France, not more than two out of twenty of the inhabitants were left alive, and the capital felt the fury of the plague, alike in the palace and the cot.

Two queens, one bishop, and great numbers of other distinguished persons, fell a sacrifice to it, and more than 500 a day died in the Hotel Dieu, under the faithful care of the sisters of charity, whose disinterested courage, in this age of horror, displayed the most beautiful traits of human virtue. For although they lost their lives, evidently from contagion, and their numbers were several times renewed, there was still no want of fresh candidates, who, strangers to the unchristian fear of death, piously devoted themselves to their holy calling.

The churchyards were soon unable to contain the dead, and many houses, left without inhabitants, fell to ruins.

In Avignon, the Pope found it necessary to consecrate the Rhone, that bodies might be thrown into the river without delay, as the churchyards would no longer hold them; so likewise, in all populous cities, extraordinary measures were adopted, in order speedily to dispose of the dead. In Vienna, where for some time 1,200 inhabitants died daily, the interment of corpses in the churchyards and within the churches was forthwith prohibited; and the dead were then arranged in layers, by thousands, in six large pits outside the city, as had already been done in Cairo and Paris. Yet, still many were secretly buried; for at all times the people are attached to the consecrated cemeteries of their dead, and will not renounce the customary mode of interment.

In many places it was rumoured that plague patients were buried alive, as may sometimes happen through senseless alarm and indecent haste; and thus the horror of the distressed people was everywhere increased. In Erfurt, after the churchyards were filled, 12,000 corpses were thrown into eleven great pits; and the like might, more or less exactly, be stated with respect to all the larger cities. Funeral ceremonies, the last consolation of the survivors, were everywhere impracticable.

In all Germany, according to a probable calculation, there seem to have died only 1,244,434 inhabitants; this country, however, was more spared than others: Italy, on the contrary, was most severely visited. It is said to have lost half its inhabitants; and this account is rendered credible from the immense losses of individual cities and provinces: for in Sardinia and Corsica, according to the account of the distinguished Florentine, John Villani, who was himself carried off by the Black Plague, scarcely a third part of the population remained alive; and it is related of the Venetians, that they engaged ships at a high rate to retreat to the islands; so that after the plague had carried off three-fourths of her inhabitants, that proud city was left forlorn and desolate. In Padua, after the cessation of the plague, two-thirds of the inhabitants were wanting; and in Florence it was prohibited to publish the numbers of dead, and to toll the bells at their funerals, in order that the living might not abandon themselves to despair.

We have more exact accounts of England; most of the great cities suffered incredible losses; above all, Yarmouth, in which 7,052 died; Bristol, Oxford, Norwich, Leicester, York, and London, where in one burial ground alone, there were interred upwards of 50,000 corpses, arranged in layers, in large pits. It is said that in the whole country scarcely a tenth part remained alive; but this estimate is evidently too high. Smaller losses were sufficient to cause those convulsions, whose consequences were felt for some centuries, in a false impulse given to civil life, and whose indirect influence, unknown to the English, has perhaps extended even to modern times.

Morals were deteriorated everywhere, and the service of God was in a great measure laid aside; for, in many places, the churches were deserted, being bereft of their priests. The instruction of the people was impeded; covetousness became general; and when tranquillity was restored, the great increase of lawyers was astonishing, to whom the endless disputes regarding inheritances offered a rich harvest. The want of priests too, throughout the country, operated very detrimentally upon the people (the lower classes being most exposed to the ravages of the plague, whilst the houses of the nobility were, in proportion, much more spared), and it was no compensation that whole bands of ignorant laymen, who had lost their wives during the pestilence, crowded into the monastic orders, that they might participate in the respectability of the priesthood, and in the rich heritages which fell in to the Church from all quarters. The sittings of Parliament, of the King's Bench, and of most of the other courts, were suspended as long as the malady raged. The laws of peace availed not during the dominion of death. Pope Clement took advantage of this state of disorder to adjust the bloody quarrel between

Edward III and Philip VI; yet he only succeeded during the period that the plague commanded peace. Philip's death (1350) annulled all treaties; and it is related that Edward, with other troops indeed, but with the same leaders and knights, again took the field. Ireland was much less heavily visited that England. The disease seems to have scarcely reached the mountainous districts of that kingdom; and Scotland too would perhaps have remained free, had not the Scots availed themselves of the discomfiture of the English to make an irruption into their territory, which terminated in the destruction of their army, by the plague and by the sword, and the extension of the pestilence, through those who escaped, over the whole country.

At the commencement, there was in England a superabundance of all the necessaries of life; but the plague, which seemed then to be the sole disease, was soon accompanied by a fatal murrain among the cattle. Wandering about without herdsmen, they fell by thousands; and, as has likewise been observed in Africa, the birds and beasts of prey are said not to have touched them. Of what nature this murrain may have been, can no more be determined, than whether it originated from communication with plague patients, or from other causes; but thus much is certain, that it did not break out until after the commencement of the Black Death. In consequence of this murrain, and the impossibility of removing the corn from the fields, there was everywhere a great rise in the price of food, which to many was inexplicable, because the harvest had been plentiful; by others it was attributed to the wicked designs of the labourers and dealers; but it really had its foundation in the actual deficiency arising from circumstances by which individual classes at all times endeavour to profit. For a whole year, until it terminated in August, 1349, the Black Plague prevailed in this beautiful island, and everywhere poisoned the springs of comfort and prosperity.

In other countries, it generally lasted only half a year, but returned frequently in individual places; on which account, some, without sufficient proof, assigned to it a period of seven years.

Spain was uninterruptedly ravaged by the Black Plague till after the year 1350, to which the frequent internal feuds and the wars with the Moors not a little contributed. Alphonso XI., whose passion for war carried him too far, died of it at the siege of Gibraltar, on the 26th of March, 1350. He was the only king in Europe who fell a sacrifice to it; but even before this period, innumerable families had been thrown into affliction. The mortality seems otherwise to have been smaller in Spain than in Italy, and about as considerable as in France.

The whole period during which the Black Plague raged with destructive violence in Europe was, with the exception of Russia, from the year 1347 to 1350. The plagues which in the sequel often returned until the year 1383, we do not consider as belonging to "the Great Mortality." They were rather common pestilences, without inflammation of the lungs, such as in former times, and in the following centuries, were excited by the matter of contagion everywhere existing, and which, on every favourable occasion, gained ground anew, as is usually the case with this frightful disease.

The concourse of large bodies of people was especially dangerous; and thus the premature celebration of the Jubilee to which Clement VI. cited the faithful to Rome (1350) during the great epidemic, caused a new eruption of the plague, from which it is said that scarcely one in a hundred of the pilgrims escaped.

Italy was, in consequence, depopulated anew; and those who returned, spread poison and corruption of morals in all directions. It is therefore the less apparent how that Pope, who was in general so wise and considerate, and who knew how to pursue the path of reason and humanity under the most difficult circumstances, should have been led to adopt a measure so injurious; since he himself was so convinced of the salutary effect of seclusion, that during the plague in Avignon he kept up constant fires, and suffered no one to approach him; and in other respects gave such orders as averted, or alleviated, much misery.

The changes which occurred about this period in the north of Europe are sufficiently memorable to claim a few moments' attention. In Sweden two princes died – Haken and Knut, half-brothers of King Magnus; and in Westgothland alone, 466 priests. The inhabitants of Iceland and Greenland found in the coldness of their inhospitable climate no protection against the southern enemy who had penetrated to them from happier countries. The plague caused great havoc among them. Nature made no allowance for their constant warfare with the elements, and the parsimony with which she had meted out to them the enjoyments of life. In Denmark and Norway, however, people were so occupied with their own misery, that the accustomed voyages to Greenland ceased. Towering icebergs formed at the same time on the coast of East Greenland, in consequence of the general concussion of the earth's organism; and no mortal, from that time forward, has ever seen that shore or its inhabitants.

It has been observed above, that in Russia the Black Plague did not break out until 1351, after it had already passed through the south and north of Europe. In this country also, the mortality was extraordinarily great; and the same scenes of affliction and despair were exhibited, as had occurred in those nations which had already passed the ordeal: the same mode of burial - the same horrible certainty of death - the same torpor and depression of spirits. The wealthy abandoned their treasures, and gave their villages and estates to the churches and monasteries; this being, according to the notions of the age, the surest way of securing the favour of Heaven and the forgiveness of past sins. In Russia, too, the voice of nature was silenced by fear and horror. In the hour of danger, fathers and mothers deserted their children, and children their parents.

Of all the estimates of the number of lives lost in Europe, the most probable is, that altogether a fourth part of the inhabitants were carried off. Now, if Europe at present contain 210,000,000 inhabitants, the population, not to take a higher estimate, which might easily by justified, amounted to at least 105,000,000 in the sixteenth century.

It may therefore be assumed, without exaggeration, that Europe lost during the Black Death 25,000,000 of inhabitants.

That her nations could so quickly overcome such a fearful concussion in their external circumstances, and, in general, without retrograding more than they actually did, could so develop their energies in the following century, is a most convincing proof of the indestructibility of human society as a whole. To assume, however, that it did not suffer any essential change internally, because in appearance everything remained as before, is inconsistent with a just view of cause and effect. Many historians seem to have adopted such an opinion; accustomed, as usual, to judge of the moral condition of the people solely according to the vicissitudes of earthly power, the events of battles, and the influence of religion, but to pass over with indifference the great phenomena of nature, which modify, not only the surface of the earth, but also the human mind. Hence, most of them have touched but superficially on the "Great Mortality" of the fourteenth century. We, for our parts, are convinced that in the history of the world the Black Death is one of the most important events which have prepared the way for the present state of Europe.

He who studies the human mind with attention, and forms a deliberate judgment on the intellectual powers which set people and States in motion, may perhaps find some proofs of this assertion in the following observations: – at that time, the advancement of the hierarchy was, in most countries, extraordinary; for the Church acquired treasures and large properties in land, even to a greater extent than after the Crusades; but experience has demonstrated that such a state of things is ruinous to the people, and causes them to retrograde, as was evinced on this occasion.

After the cessation of the Black Plague, a greater fecundity in women was everywhere remarkable – a grand phenomenon, which, from its occurrence after every destructive pestilence, proves to conviction, if any occurrence can do so, the prevalence of a higher power in the direction of general organic life. Marriages were, almost without exception, prolific; and double and triple births were more frequent than at other times; under which head, we should remember the strange remark, that after the "Great Mortality" the children were said to have got fewer teeth than before; at which contemporaries were mightily shocked, and even later writers have felt surprise.

If we examine the grounds of this oft-repeated assertion, we shall find that they were astonished to see children, cut twenty, or at most, twenty-two teeth, under the supposition that a greater number had formerly fallen to their share. Some writers of authority, as, for example, the physician Savonarola, at Ferrara, who probably looked for twenty-eight teeth in children, published their opinions on this subject. Others copied from them, without seeing for themselves, as often happens in other matters which are equally evident; and thus the world believed in the miracle of an imperfection in the human body which had been caused by the Black Plague.

The people gradually consoled themselves after the sufferings which they had undergone; the dead were lamented and forgotten; and, in the stirring vicissitudes of existence, the world belonged to the living.

CHAPTER V

MORAL EFFECTS

***

The mental shock sustained by all nations during the prevalence of the Black Plague is without parallel and beyond description. In the eyes of the timorous, danger was the certain harbinger of death; many fell victims to fear on the first appearance of the distemper, and the most stouthearted lost their confidence. Thus, after reliance on the future had died away, the spiritual union which binds man to his family and his fellow-creatures was gradually dissolved. The pious closed their accounts with the world – eternity presented itself to their view – their only remaining desire was for a participation in the consolations of religion, because to them death was disarmed of its sting.

Repentance seized the transgressor, admonishing him to consecrate his remaining hours to the exercise of Christian virtues. All minds were directed to the contemplation of futurity; and children, who manifest the more elevated feelings of the soul without alloy, were frequently seen, while labouring under the plague, breathing out their spirit with prayer and songs of thanksgiving.

An awful sense of contrition seized Christians of every communion; they resolved to forsake their vices, to make restitution for past offences, before they were summoned hence, to seek reconciliation with their Maker, and to avert, by self-chastisement, the punishment due to their former sins. Human nature would be exalted, could the countless noble actions which, in times of most imminent danger, were performed in secret, be recorded for the instruction of future generations. They, however, have no influence on the course of worldly events. They are known only to silent eyewitnesses, and soon fall into oblivion. But hypocrisy, illusion, and bigotry stalk abroad undaunted; they desecrate what is noble, they pervert what is divine, to the unholy purposes of selfishness, which hurries along every good feeling in the false excitement of the age. Thus it was in the years of this plague. In the fourteenth century, the monastic system was still in its full vigour, the power of the ecclesiastical orders and brotherhoods was revered by the people, and the hierarchy was still formidable to the temporal power. It was therefore in the natural constitution of society that bigoted zeal, which in such times makes a show of public acts of penance, should avail itself of the semblance of religion. But this took place in such a manner, that unbridled, self-willed penitence, degenerated into lukewarmness, renounced obedience to the hierarchy, and prepared a fearful opposition to the Church, paralysed as it was by antiquated forms.

While all countries were filled with lamentations and woe, there first arose in Hungary, and afterwards in Germany, the Brotherhood of the Flagellants, called also the Brethren of the Cross, or Cross-bearers, who took upon themselves the repentance of the people for the sins they had committed, and offered prayers and supplications for the averting of this plague. This Order consisted chiefly of persons of the lower class, who were either actuated by sincere contrition, or who joyfully availed themselves of this pretext for idleness, and were hurried along with the tide of distracting frenzy. But as these brotherhoods gained in repute, and were welcomed by the people with veneration and enthusiasm, many nobles and ecclesiastics ranged themselves under their standard; and their bands were not unfrequently augmented by children, honourable women, and nuns; so powerfully were minds of the most opposite temperaments enslaved by this infatuation. They marched through the cities, in well-organised processions, with leaders and singers; their heads covered as far as the eyes; their look fixed on the ground, accompanied by every token of the deepest contrition and mourning. They were robed in sombre garments, with red crosses on the breast, back, and cap, and bore triple scourges, tied in three or four knots, in which points of iron were fixed. Tapers and magnificent banners of velvet and cloth of gold were carried before them; wherever they made their appearance, they were welcomed by the ringing bells, and the people flocked from all quarters to listen to their hymns and to witness their penance with devotion and tears.

In the year 1349, two hundred Flagellants first entered Strasburg, where they were received with great joy, and hospitably lodged by citizens. Above a thousand joined the brotherhood, which now assumed the appearance of a wandering tribe, and separated into two bodies, for the purpose of journeying to the north and to the south. For more than half a year, new parties arrived weekly; and on each arrival adults and children left their families to accompany them; till at length their sanctity was questioned, and the doors of houses and churches were closed against them. At Spires, two hundred boys, of twelve years of age and under, constituted themselves into a Brotherhood of the Cross, in imitation of the children who, about a hundred years before, had united, at the instigation of some fanatic monks, for the purpose of recovering the Holy Sepulchre. All the inhabitants of this town were carried away by the illusion; they conducted the strangers to their houses with songs of thanksgiving, to regale them for the night. The women embroidered banners for them, and all were anxious to augment their pomp; and at every succeeding pilgrimage their influence and reputation increased.

It was not merely some individual parts of the country that fostered them: all Germany, Hungary, Poland, Bohemia, Silesia, and Flanders, did homage to the mania; and they at length became as formidable to the secular as they were to the ecclesiastical power. The influence of this fanaticism was great and threatening, resembling the excitement which called all the inhabitants of Europe into the deserts of Syria and Palestine about two hundred and fifty years before. The appearance in itself was not novel. As far back as the eleventh century, many believers in Asia and Southern Europe afflicted themselves with the punishment of flagellation. Dominicus Loricatus, a monk of St. Croce d'Avellano, is mentioned as the master and model of this species of mortification of the flesh; which, according to the primitive notions of the Asiatic Anchorites, was deemed eminently Christian. The author of the solemn processions of the Flagellants is said to have been St. Anthony; for even in his time (1231) this kind of penance was so much in vogue, that it is recorded as an eventful circumstance in the history of the world. In 1260, the Flagellants appeared in Italy as Devoti. "When the land was polluted by vices and crimes, an unexampled spirit of remorse suddenly seized the minds of the Italians. The fear of Christ fell upon all: noble and ignoble, old and young, and even children of five years of age, marched through the streets with no covering but a scarf round the waist. They each carried a scourge of leathern thongs, which they applied to their limbs, amid sighs and tears, with such violence that the blood flowed from the wounds. Not only during the day, but even by night, and in the severest winter, they traversed the cities with burning torches and banners, in thousands and tens of thousands, headed by their priests, and prostrated themselves before the altars. They proceeded in the same manner in the villages: and the woods and mountains resounded with the voices of those whose cries were raised to God. The melancholy chaunt of the penitent alone was heard. Enemies were reconciled; men and women vied with each other in splendid works of charity, as if they dreaded that Divine Omnipotence would pronounce on them the doom of annihilation."

The pilgrimages of the Flagellants extended throughout all the province of Southern Germany, as far as Saxony, Bohemia, and Poland, and even further; but at length the priests resisted this dangerous fanaticism, without being able to extirpate the illusion, which was advantageous to the hierarchy as long as it submitted to its sway. Regnier, a hermit of Perugia, is recorded as a fanatic preacher of penitence, with whom the extravagance originated. In the year 1296 there was a great procession of the Flagellants in Strasburg; and in 1334, fourteen years before the Great Mortality, the sermon of Venturinus, a Dominican friar of Bergamo, induced above 10,000 persons to undertake a new pilgrimage. They scourged themselves in the churches, and were entertained in the market-places at the public expense. At Rome, Venturinus was derided, and banished by the Pope to the mountains of Ricondona. He patiently endured all – went to the Holy Land, and died at Smyrna, 1346. Hence we see that this fanaticism was a mania of the middle ages, which, in the year 1349, on so fearful an occasion, and while still so fresh in remembrance, needed no new founder; of whom, indeed, all the records are silent. It probably arose in many places at the same time; for the terror of death, which pervaded all nations and suddenly set such powerful impulses in motion, might easily conjure up the fanaticism of exaggerated and overpowering repentance.

The manner and proceedings of the Flagellants of the thirteenth and fourteenth centuries exactly resemble each other. But, if during the Black Plague, simple credulity came to their aid, which seized, as a consolation, the grossest delusion of religious enthusiasm, yet it is evident that the leaders must have been intimately united, and have exercised the power of a secret association. Besides, the rude band was generally under the control of men of learning, some of whom at least certainly had other objects in view independent of those which ostensibly appeared. Whoever was desirous of joining the brotherhood, was bound to remain in it thirty-four days, and to have fourpence per day at his own disposal, so that he might not be burthensome to any one; if married, he was obliged to have the sanction of his wife, and give the assurance that he was reconciled to all men. The Brothers of the Cross were not permitted to seek for free quarters, or even to enter a house without having been invited; they were forbidden to converse with females; and if they transgressed these rules, or acted without discretion, they were obliged to confess to the Superior, who sentenced them to several lashes of the scourge, by way of penance. Ecclesiastics had not, as such, any pre-eminence among them; according to their original law, which, however, was often transgressed, they could not become Masters, or take part in the Secret Councils. Penance was performed twice every day: in the morning and evening they went abroad in pairs, singing psalms amid the ringing of the bells; and when they arrived at the place of flagellation, they stripped the upper part of their bodies and put off their shoes, keeping on only a linen dress, reaching from the waist to the ankles. They then lay down in a large circle, in different positions, according to the nature of the crime: the adulterer with his face to the ground; the perjurer on one side, holding up three of his fingers, &c., and were then castigated, some more and some less, by the Master, who ordered them to rise in the words of a prescribed form. Upon this they scourged themselves, amid the singing of psalms and loud supplications for the averting of the plague, with genuflexions and other ceremonies, of which contemporary writers give various accounts; and at the same time constantly boasted of their penance, that the blood of their wounds was mingled with that of the Saviour. One of them, in conclusion, stoop up to read a letter, which it was pretended an angel had brought from heaven to St. Peter's Church, at Jerusalem, stating that Christ, who was sore displeased at the sins of man, had granted, at the intercession of the Holy Virgin and of the angels, that all who should wander about for thirty-four days and scourge themselves, should be partakers of the Divine grace. This scene caused as great a commotion among the believers as the finding of the holy spear once did at Antioch; and if any among the clergy inquired who had sealed the letter, he was boldly answered, the same who had sealed the Gospel!

All this had so powerful an effect, that the Church was in considerable danger; for the Flagellants gained more credit than the priests, from whom they so entirely withdrew themselves, that they even absolved each other. Besides, they everywhere took possession of the churches, and their new songs, which went from mouth to mouth, operated strongly on the minds of the people. Great enthusiasm and originally pious feelings are clearly distinguishable in these hymns, and especially in the chief psalm of the Cross-bearers, which is still extant, and which was sung all over Germany in different dialects, and is probably of a more ancient date. Degeneracy, however, soon crept in; crimes were everywhere committed; and there was no energetic man capable of directing the individual excitement to purer objects, even had an effectual resistance to the tottering Church been at that early period seasonable, and had it been possible to restrain the fanaticism. The Flagellants sometimes undertook to make trial of their power of working miracles; as in Strasburg, where they attempted, in their own circle, to resuscitate a dead child: they, however, failed, and their unskilfulness did them much harm, though they succeeded here and there in maintaining some confidence in their holy calling, by pretending to have the power of casting out evil spirits.

The Brotherhood of the Cross announced that the pilgrimage of the Flagellants was to continue for a space of thirty-four years; and many of the Masters had doubtless determined to form a lasting league against the Church; but they had gone too far. So early as the first year of their establishment, the general indignation set bounds to their intrigues: so that the strict measures adopted by the Emperor Charles IV., and Pope Clement, who, throughout the whole of this fearful period, manifested prudence and noble-mindedness, and conducted himself in a manner every way worthy of his high station, were easily put into execution.

The Sorbonne, at Paris, and the Emperor Charles, had already applied to the Holy See for assistance against these formidable and heretical excesses, which had well-nigh destroyed the influence of the clergy in every place; when a hundred of the Brotherhood of the Cross arrived at Avignon from Basle, and desired admission. The Pope, regardless of the intercession of several cardinals, interdicted their public penance, which he had not authorised; and, on pain of excommunication, prohibited throughout Christendom the continuance of these pilgrimages. Philip VI., supported by the condemnatory judgment of the Sorbonne, forbade their reception in France. Manfred, King of Sicily, at the same time threatened them with punishment by death; and in the East they were withstood by several bishops, among whom was Janussius, of Gnesen, and Preczlaw, of Breslau, who condemned to death one of their Masters, formerly a deacon; and, in conformity with the barbarity of the times, had him publicly burnt. In Westphalia, where so shortly before they had venerated the Brothers of the Cross, they now persecuted them with relentless severity; and in the Mark, as well as in all the other countries of Germany, they pursued them as if they had been the authors of every misfortune.

The processions of the Brotherhood of the Cross undoubtedly promoted the spreading of the plague; and it is evident that the gloomy fanaticism which gave rise to them would infuse a new poison into the already desponding minds of the people.

Still, however, all this was within the bounds of barbarous enthusiasm; but horrible were the persecutions of the Jews, which were committed in most countries, with even greater exasperation than in the twelfth century, during the first Crusades. In every destructive pestilence the common people at first attribute the mortality to poison. No instruction avails; the supposed testimony of their eyesight is to them a proof, and they authoritatively demand the victims of their rage. On whom, then, was it so likely to fall as on the Jews, the usurers and the strangers who lived at enmity with the Christians? They were everywhere suspected of having poisoned the wells or infected the air. They alone were considered as having brought this fearful mortality upon the Christians. They were, in consequence, pursued with merciless cruelty; and either indiscriminately given up to the fury of the populace, or sentenced by sanguinary tribunals, which, with all the forms of the law, ordered them to be burnt alive. In times like these, much is indeed said of guilt and innocence; but hatred and revenge bear down all discrimination, and the smallest probability magnifies suspicion into certainty. These bloody scenes, which disgraced Europe in the fourteenth century, are a counterpart to a similar mania of the age, which was manifested in the persecutions of witches and sorcerers; and, like these, they prove that enthusiasm, associated with hatred, and leagued with the baser passions, may work more powerfully upon whole nations than religion and legal order; nay, that it even knows how to profit by the authority of both, in order the more surely to satiate with blood the sword of long-suppressed revenge.

The persecution of the Jews commenced in September and October, 1348, at Chillon, on the Lake of Geneva, where the first criminal proceedings were instituted against them, after they had long before been accused by the people of poisoning the wells; similar scenes followed in Bern and Freyburg, in January, 1349. Under the influence of excruciating suffering, the tortured Jews confessed themselves guilty of the crime imputed to them; and it being affirmed that poison had in fact been found in a well at Zoffingen, this was deemed a sufficient proof to convince the world; and the persecution of the abhorred culprits thus appeared justifiable. Now, though we can take as little exception at these proceedings as at the multifarious confessions of witches, because the interrogatories of the fanatical and sanguinary tribunals were so complicated, that by means of the rack the required answer must inevitably be obtained; and it is, besides, conformable to human nature that crimes which are in everybody's mouth may, in the end, be actually committed by some, either from wantonness, revenge, or desperate exasperation: yet crimes and accusations are, under circumstances like these, merely the offspring of a revengeful, frenzied spirit in the people; and the accusers, according to the fundamental principles of morality, which are the same in every age, are the more guilty transgressors.

Already in the autumn of 1348 a dreadful panic, caused by this supposed empoisonment, seized all nations; in Germany especially the springs and wells were built over, that nobody might drink of them or employ their contents for culinary purposes; and for a long time the inhabitants of numerous towns and villages used only river and rain water. The city gates were also guarded with the greatest caution: only confidential persons were admitted; and if medicine or any other article, which might be supposed to be poisonous, was found in the possession of a stranger – and it was natural that some should have these things by them for their private use – they were forced to swallow a portion of it. By this trying state of privation, distrust, and suspicion, the hatred against the supposed poisoners became greatly increased, and often broke out in popular commotions, which only served still further to infuriate the wildest passions. The noble and the mean fearlessly bound themselves by an oath to extirpate the Jews by fire and sword, and to snatch them from their protectors, of whom the number was so small, that throughout all Germany but few places can be mentioned where these unfortunate people were not regarded as outlaws and martyred and burnt. Solemn summonses were issued from Bern to the towns of Basle, Freyburg in the Breisgau, and Strasburg, to pursue the Jews as poisoners. The burgomasters and senators, indeed, opposed this requisition; but in Basle the populace obliged them to bind themselves by an oath to burn the Jews, and to forbid persons of that community from entering their city for the space of two hundred years. Upon this all the Jews in Basle, whose number could not have been inconsiderable, were enclosed in a wooden building, constructed for the purpose, and burnt together with it, upon the mere outcry of the people, without sentence or trial, which, indeed, would have availed them nothing. Soon after the same thing took place at Freyburg. A regular Diet was held at Bennefeld, in Alsace, where the bishops, lords, and barons, as also deputies of the counties and towns, consulted how they should proceed with regard to the Jews; and when the deputies of Strasburg – not indeed the bishop of this town, who proved himself a violent fanatic – spoke in favour of the persecuted, as nothing criminal was substantiated against them, a great outcry was raised, and it was vehemently asked, why, if so, they had covered their wells and removed their buckets. A sanguinary decree was resolved upon, of which the populace, who obeyed the call of the nobles and superior clergy, became but the too willing executioners. Wherever the Jews were not burnt, they were at least banished; and so being compelled to wander about, they fell into the hands of the country people, who, without humanity, and regardless of all laws, persecuted them with fire and sword. At Spires, the Jews, driven to despair, assembled in their own habitations, which they set on fire, and thus consumed themselves with their families. The few that remained were forced to submit to baptism; while the dead bodies of the murdered, which lay about the streets, were put into empty wine-casks and rolled into the Rhine, lest they should infect the air. The mob was forbidden to enter the ruins of the habitations that were burnt in the Jewish quarter; for the senate itself caused search to be made for the treasure, which is said to have been very considerable. At Strasburg two thousand Jews were burnt alive in their own burial-ground, where a large scaffold had been erected: a few who promised to embrace Christianity were spared, and their children taken from the pile. The youth and beauty of several females also excited some commiseration, and they were snatched from death against their will; many, however, who forcibly made their escape from the flames were murdered in the streets.

The senate ordered all pledges and bonds to be returned to the debtors, and divided the money among the workpeople. Many, however, refused to accept the base price of blood, and, indignant at the scenes of bloodthirsty avarice, which made the infuriated multitude forget that the plague was raging around them, presented it to monasteries, in conformity with the advice of their confessors. In all the countries on the Rhine, these cruelties continued to be perpetrated during the succeeding months; and after quiet was in some degree restored, the people thought to render an acceptable service to God, by taking the bricks of the destroyed dwellings, and the tombstones of the Jews, to repair churches and to erect belfries.

In Mayence alone, 12,000 Jews are said to have been put to a cruel death. The Flagellants entered that place in August; the Jews, on this occasion, fell out with the Christians and killed several; but when they saw their inability to withstand the increasing superiority of their enemies, and that nothing could save them from destruction, they consumed themselves and their families by setting fire to their dwellings. Thus also, in other places, the entry of the Flagellants gave rise to scenes of slaughter; and as thirst for blood was everywhere combined with an unbridled spirit of proselytism, a fanatic zeal arose among the Jews to perish as martyrs to their ancient religion. And how was it possible that they could from the heart embrace Christianity, when its precepts were never more outrageously violated? At Eslingen the whole Jewish community burned themselves in their synagogue, and mothers were often seen throwing their children on the pile, to prevent their being baptised, and then precipitating themselves into the flames. In short, whatever deeds fanaticism, revenge, avarice and desperation, in fearful combination, could instigate mankind to perform, – and where in such a case is the limit? – were executed in the year 1349 throughout Germany, Italy, and France, with impunity, and in the eyes of all the world. It seemed as if the plague gave rise to scandalous acts and frantic tumults, not to mourning and grief; and the greater part of those who, by their education and rank, were called upon to raise the voice of reason, themselves led on the savage mob to murder and to plunder. Almost all the Jews who saved their lives by baptism were afterwards burnt at different times; for they continued to be accused of poisoning the water and the air.

Christians also, whom philanthropy or gain had induced to offer them protection, were put on the rack and executed with them. Many Jews who had embraced Christianity repented of their apostacy, and, returning to their former faith, sealed it with their death.

The humanity and prudence of Clement VI. must, on this occasion, also be mentioned to his honour; but even the highest ecclesiastical power was insufficient to restrain the unbridled fury of the people. He not only protected the Jews at Avignon, as far as lay in his power, but also issued two bulls, in which he declared them innocent; and admonished all Christians, though without success, to cease from such groundless persecutions. The Emperor Charles IV. was also favourable to them, and sought to avert their destruction wherever he could; but he dared not draw the sword of justice, and even found himself obliged to yield to the selfishness of the Bohemian nobles, who were unwilling to forego so favourable an opportunity of releasing themselves from their Jewish creditors, under favour of an imperial mandate. Duke Albert of Austria burnt and pillaged those of his cities which had persecuted the Jews – a vain and inhuman proceeding, which, moreover, is not exempt from the suspicion of covetousness; yet he was unable, in his own fortress of Kyberg, to protect some hundreds of Jews, who had been received there, from being barbarously burnt by the inhabitants. Several other princes and counts, among whom was Ruprecht von der Pfalz, took the Jews under their protection, on the payment of large sums: in consequence of which they were called "Jew-masters," and were in danger of being attacked by the populace and by their powerful neighbours. These persecuted and ill-used people, except indeed where humane individuals took compassion on them at their own peril, or when they could command riches to purchase protection, had no place of refuge left but the distant country of Lithuania, where Boleslav V., Duke of Poland (1227-1279) had before granted them liberty of conscience; and King Casimir the Great (1333-1370), yielding to the entreaties of Esther, a favourite Jewess, received them, and granted them further protection; on which account, that country is still inhabited by a great number of Jews, who by their secluded habits have, more than any people in Europe, retained the manners of the Middle Ages.

But to return to the fearful accusations against the Jews; it was reported in all Europe that they were in connection with secret superiors in Toledo, to whose decrees they were subject, and from whom they had received commands respecting the coining of base money, poisoning, the murder of Christian children, &c; that they received the poison by sea from remote parts, and also prepared it themselves from spiders, owls, and other venomous animals; but, in order that their secret might not be discovered, that it was known only to their Rabbis and rich men. Apparently there were but few who did not consider this extravagant accusation well founded; indeed, in many writings of the fourteenth century, we find great acrimony with regard to the suspected poison-mixers, which plainly demonstrates the prejudice existing against them. Unhappily, after the confessions of the first victims in Switzerland, the rack extorted similar ones in various places. Some even acknowledged having received poisonous powder in bags, and injunctions from Toledo, by secret messengers. Bags of this description were also often found in wells, though it was not unfrequently discovered that the Christians themselves had thrown them in; probably to give occasion to murder and pillage; similar instances of which may be found in the persecutions of the witches.

This picture needs no additions. A lively image of the Black Plague, and of the moral evil which followed in its train, will vividly represent itself to him who is acquainted with nature and the constitution of society. Almost the only credible accounts of the manner of living, and of the ruin which occurred in private life during this pestilence, are from Italy; and these may enable us to form a just estimate of the general state of families in Europe, taking into consideration what is peculiar in the manners of each country.

"When the evil had become universal" (speaking of Florence), "the hearts of all the inhabitants were closed to feelings of humanity. They fled from the sick and all that belonged to them, hoping by these means to save themselves. Others shut themselves up in their houses, with their wives, their children and households, living on the most costly food, but carefully avoiding all excess. None were allowed access to them; no intelligence of death or sickness was permitted to reach their ears; and they spent their time in singing and music, and other pastimes. Others, on the contrary, considered eating and drinking to excess, amusements of all descriptions, the indulgence of every gratification, and an indifference to what was passing around them, as the best medicine, and acted accordingly. They wandered day and night from one tavern to another, and feasted without moderation or bounds. In this way they endeavoured to avoid all contact with the sick, and abandoned their houses and property to chance, like men whose death-knell had already tolled.

"Amid this general lamentation and woe, the influence and authority of every law, human and divine, vanished. Most of those who were in office had been carried off by the plague, or lay sick, or had lost so many members of their family, that they were unable to attend to their duties; so that thenceforth every one acted as he thought proper. Others in their mode of living chose a middle course. They ate and drank what they pleased, and walked abroad, carrying odoriferous flowers, herbs, or spices, which they smelt to from time to time, in order to invigorate the brain, and to avert the baneful influence of the air, infected by the sick and by the innumerable corpses of those who had died of the plague. Others carried their precaution still further, and thought the surest way to escape death was by flight. They therefore left the city; women as well as men abandoning their dwellings and their relations, and retiring into the country. But of these also many were carried off, most of them alone and deserted by all the world, themselves having previously set the example. Thus it was that one citizen fled from another – a neighbour from his neighbours – a relation from his relations; and in the end, so completely had terror extinguished every kindlier feeling, that the brother forsook the brother – the sister the sister – the wife her husband; and at last, even the parent his own offspring, and abandoned them, unvisited and unsoothed, to their fate. Those, therefore, that stood in need of assistance fell a prey to greedy attendants, who, for an exorbitant recompense, merely handed the sick their food and medicine, remained with them in their last moments, and then not unfrequently became themselves victims to their avarice and lived not to enjoy their extorted gain. Propriety and decorum were extinguished among the helpless sick. Females of rank seemed to forget their natural bashfulness, and committed the care of their persons, indiscriminately, to men and women of the lowest order. No longer were women, relatives or friends, found in the house of mourning, to share the grief of the survivors – no longer was the corpse accompanied to the grave by neighbours and a numerous train of priests, carrying wax tapers and singing psalms, nor was it borne along by other citizens of equal rank. Many breathed their last without a friend to soothe their dying pillow; and few indeed were they who departed amid the lamentations and tears of their friends and kindred. Instead of sorrow and mourning, appeared indifference, frivolity and mirth; this being considered, especially by the females, as conducive to health. Seldom was the body followed by even ten or twelve attendants; and instead of the usual bearers and sextons, mercenaries of the lowest of the populace undertook the office for the sake of gain; and accompanied by only a few priests, and often without a single taper, it was borne to the very nearest church, and lowered into the grave that was not already too full to receive it. Among the middling classes, and especially among the poor, the misery was still greater. Poverty or negligence induced most of these to remain in their dwellings, or in the immediate neighbourhood; and thus they fell by thousands; and many ended their lives in the streets by day and by night. The stench of putrefying corpses was often the first indication to their neighbours that more deaths had occurred. The survivors, to preserve themselves from infection, generally had the bodies taken out of the houses and laid before the doors; where the early morning found them in heaps, exposed to the affrighted gaze of the passing stranger. It was no longer possible to have a bier for every corpse – three or four were generally laid together – husband and wife, father and mother, with two or three children, were frequently borne to the grave on the same bier; and it often happened that two priests would accompany a coffin, bearing the cross before it, and be joined on the way by several other funerals; so that instead of one, there were five or six bodies for interment."

Thus far Boccacio. On the conduct of the priests, another contemporary observes: "In large and small towns they had withdrawn themselves through fear, leaving the performance of ecclesiastical duties to the few who were found courageous and faithful enough to undertake them." But we ought not on that account to throw more blame on them than on others; for we find proofs of the same timidity and heartlessness in every class. During the prevalence of the Black Plague, the charitable orders conducted themselves admirably, and did as much good as can be done by individual bodies in times of great misery and destruction, when compassion, courage, and the nobler feelings are found but in the few, while cowardice, selfishness and ill-will, with the baser passions in their train, assert the supremacy. In place of virtue which had been driven from the earth, wickedness everywhere reared her rebellious standard, and succeeding generations were consigned to the dominion of her baleful tyranny.

CHAPTER VI

PHYSICIANS

***

If we now turn to the medical talent which encountered the "Great Mortality," the Middle Ages must stand excused, since even the moderns are of opinion that the art of medicine is not able to cope with the Oriental plague, and can afford deliverance from it only under particularly favourable circumstances. We must bear in mind, also, that human science and art appear particularly weak in great pestilences, because they have to contend with the powers of nature, of which they have no knowledge; and which, if they had been, or could be, comprehended in their collective effects, would remain uncontrollable by them, principally on account of the disordered condition of human society. Moreover, every new plague has its peculiarities, which are the less easily discovered on first view because, during its ravages, fear and consternation humble the proud spirit.

The physicians of the fourteenth century, during the Black Death, did what human intellect could do in the actual condition of the healing art; and their knowledge of the disease was by no means despicable. They, like the rest of mankind, have indulged in prejudices, and defended them, perhaps, with too much obstinacy: some of these, however, were founded on the mode of thinking of the age, and passed current in those days as established truths; others continue to exist to the present hour.

Their successors in the nineteenth century ought not therefore to vaunt too highly the pre-eminence of their knowledge, for they too will be subjected to the severe judgment of posterity – they too will, with reason, be accused of human weakness and want of foresight.

The medical faculty of Paris, the most celebrated of the fourteenth century, were commissioned to deliver their opinion on the causes of the Black Plague, and to furnish some appropriate regulations with regard to living during its prevalence. This document is sufficiently remarkable to find a place here.

"We, the Members of the College of Physicians of Paris, have, after mature consideration and consultation on the present mortality, collected the advice of our old masters in the art, and intend to make known the causes of this pestilence more clearly than could be done according to the rules and principles of astrology and natural science; we, therefore, declare as follows:–

> "It is known that in India and the vicinity of the Great Sea, the constellations which combated the rays of the sun, and the warmth of the heavenly fire, exerted their power especially against that sea, and struggled violently with its waters. (Hence vapours often originate

which envelop the sun, and convert his light into darkness.) These vapours alternately rose and fell for twenty-eight days; but, at last, sun and fire acted so powerfully upon the sea that they attracted a great portion of it to themselves, and the waters of the ocean arose in the form of vapour; thereby the waters were in some parts so corrupted that the fish which they contained died. These corrupted waters, however, the heat of the sun could not consume, neither could other wholesome water, hail or snow and dew, originate therefrom. On the contrary, this vapour spread itself through the air in many places on the earth, and enveloped them in fog.

"Such was the case all over Arabia, in a part of India, in Crete, in the plains and valleys of Macedonia, in Hungary, Albania, and Sicily. Should the same thing occur in Sardinia, not a man will be left alive, and the like will continue so long as the sun remains in the sign of Leo, on all the islands and adjoining countries to which this corrupted sea-wind extends, or has already extended, from India. If the inhabitants of those parts do not employ and adhere to the following or similar means and precepts, we announce to them inevitable death, except the grace of Christ preserve their lives.

"We are of opinion that the constellations, with the aid of nature, strive by virtue of their Divine might, to protect and heal the human race; and to this end, in union with the rays of the sun, acting through the power of fire, endeavour to break through the mist. Accordingly, within the next ten days, and until the 17th of the ensuing month of July, this mist will be converted into a stinking deleterious rain, whereby the air will be much purified. Now, as soon as this rain shall announce itself by thunder or hail, every one of you should protect himself from the air; and, as well before as after the rain, kindle a large fire of vinewood, green laurel, or other green wood; wormwood and camomile should also be burnt in great quantity in the marketplaces, in other densely inhabited localities, and in the houses. Until the earth is again completely dry, and for three days afterwards, no one ought to go abroad in the fields. During this time the diet should be simple, and people should be cautious in avoiding exposure in the cool of the evening, at night, and in the morning. Poultry and waterfowl, young pork, old beef, and fat meat in general, should not be eaten; but, on the contrary, meat of a proper age, of a warm and dry, but on no account of a heating and exciting nature. Broth should be taken, seasoned with ground pepper, ginger, and cloves, especially by those who are accustomed to live temperately, and are yet choice in their diet. Sleep in the daytime is detrimental; it should be taken at night until sunrise, or somewhat longer. At breakfast one should drink little; supper should be taken an hour before sunset, when more may be drunk than in the morning. Clear light wine, mixed with a fifth or six part of water, should be used as a beverage. Dried or fresh fruits, with wine, are not injurious, but highly so without it. beetroot and other vegetables, whether eaten pickled or fresh, are hurtful; on the contrary, spicy pot-herbs, as sage or rosemary, are wholesome. Cold, moist, watery food in is general prejudicial. Going out at night, and even until three o'clock in the morning, is dangerous, on account of dew. Only small river fish should be used. Too much exercise is hurtful. The body should be kept warmer than usual, and thus protected from moisture and cold. rainwater must not be employed in cooking, and every one should guard against exposure to wet weather. If it rain, a little fine treacle should be taken after dinner. Fat people should not sit in the sunshine. Good

> clear wine should be selected and drunk often, but in small quantities, by day. Olive oil as an article of food is fatal. Equally injurious are fasting and excessive abstemiousness, anxiety of mind, anger, and immoderate drinking. Young people, in autumn especially, must abstain from all these things if they do not wish to run a risk of dying of dysentery. In order to keep the body properly open, an enema, or some other simple means, should be employed when necessary. Bathing is injurious. Men must preserve chastity as they value their lives. Every one should impress this on his recollection, but especially those who reside on the coast, or upon an island into which the noxious wind has penetrated."

On what occasion these strange precepts were delivered can no longer be ascertained, even if it were an object to know it. It must be acknowledged, however, that they do not redound to the credit either of the faculty of Paris, or of the fourteenth century in general. This famous faculty found themselves under the painful necessity of being wise at command, and of firing a pointblank shot of erudition at an enemy who enveloped himself in a dark mist, of the nature of which they had no conception. In concealing their ignorance by authoritative assertions, they suffered themselves, therefore, to be misled; and while endeavouring to appear to the world with eclat, only betrayed to the intelligent their lamentable weakness. Now some might suppose that, in the condition of the sciences of the fourteenth century, no intelligent physicians existed; but this is altogether at variance with the laws of human advancement, and is contradicted by history. The real knowledge of an age is shown only in the archives of its literature. Here alone the genius of truth speaks audibly here alone men of talent deposit the results of their experience and reflection without vanity or a selfish object. There is no ground for believing that in the fourteenth century men of this kind were publicly questioned regarding their views; and it is, therefore, the more necessary that impartial history should take up their cause, and do justice to their merits.

The first notice on this subject is due to a very celebrated teacher in Perugia, Gentilis of Foligno, who, on the 18th of June, 1348, fell a sacrifice to the plague, in the faithful discharge of his duty. Attached to Arabian doctrines, and to the universally respected Galen, he, in common with all his contemporaries, believed in a putrid corruption of the blood in the lungs and in the heart, which was occasioned by the pestilential atmosphere, and was forthwith communicated to the whole body. He thought, therefore, that everything depended upon a sufficient purification of the air, by means of large blazing fires of odoriferous wood, in the vicinity of the healthy as well as of the sick, and also upon an appropriate manner of living, so that the putridity might not overpower the diseased. In conformity with notions derived from the ancients, he depended upon bleeding and purging, at the commencement of the attack, for the purpose of purification; ordered the healthy to wash themselves frequently with vinegar or wine, to sprinkle their dwellings with vinegar, and to smell often to camphor, or other volatile substances. Hereupon he gave, after the Arabian fashion, detailed rules, with an abundance of different medicines, of whose healing powers wonderful things were believed. He had little stress upon super-lunar influences, so far as respected the malady itself; on which account, he did not enter into the great controversies of the astrologers, but always kept in view, as an object of medical attention, the corruption of the blood in the lungs and heart. He believed in a progressive infection from country to country, according to the notions of the present day; and the contagious power of the disease, even in the vicinity of those affected by plague, was, in his opinion, beyond all doubt. On this point intelligent contemporaries were all agreed; and, in truth, it required no great genius to be convinced of so palpable a fact.

Besides, correct notions of contagion have descended from remote antiquity, and were maintained unchanged in the fourteenth century. So far back as the age of Plato a knowledge of the contagious power of malignant inflammations of the eye, of which also no physician of the Middle Ages entertained a doubt, was general among the people; yet in modern times surgeons have filled volumes with partial controversies on this subject. The whole language of antiquity has adapted itself to the notions of the people respecting the contagion of pestilential diseases; and their terms were, beyond comparison, more expressive than those in use among the moderns.

Arrangements for the protection of the healthy against contagious diseases, the necessity of which is shown from these notions, were regarded by the ancients as useful; and by man, whose circumstances permitted it, were carried into effect in their houses. Even a total separation of the sick from the healthy, that indispensable means of protection against infection by contact, was proposed by physicians of the second century after Christ, in order to check the spreading of leprosy. But it was decidedly opposed, because, as it was alleged, the healing art ought not to be guilty of such harshness. This mildness of the ancients, in whose manner of thinking inhumanity was so often and so undisguisedly conspicuous, might excite surprise if it were anything more than apparent. The true ground of the neglect of public protection against pestilential diseases lay in the general notion and constitution of human society – it lay in the disregard of human life, of which the great nations of antiquity have given proofs in every page of their history. Let it not be supposed that they wanted knowledge respecting the propagation of contagious diseases. On the contrary, they were as well informed on this subject as the modern; but this was shown where individual property, not where human life, on the grand scale was to be protected. Hence the ancients made a general practice of arresting the progress of murrains among cattle by a separation of the diseased from the healthy. Their herds alone enjoyed that protection which they held it impracticable to extend to human society, because they had no wish to do so. That the governments in the fourteenth century were not yet so far advanced as to put into practice general regulations for checking the plague needs no especial proof. Physicians could, therefore, only advise public purifications of the air by means of large fires, as had often been practised in ancient times; and they were obliged to leave it to individual families either to seek safety in flight, or to shut themselves up in their dwellings, a method which answers in common plagues, but which here afforded no complete security, because such was the fury of the disease when it was at its height, that the atmosphere of whole cities was penetrated by the infection.

Of the astral influence which was considered to have originated the "Great Mortality," physicians and learned men were as completely convinced as of the fact of its reality. A grand conjunction of the three superior planets, Saturn, Jupiter, and Mars, in the sign of Aquarius, which took place, according to Guy de Chauliac, on the 24th of March, 1345, was generally received as its principal cause. In fixing the day, this physician, who was deeply versed in astrology, did not agree with others; whereupon there arose various disputations, of weight in that age, but of none in ours. People, however, agree in this – that conjunctions of the planets infallibly prognosticated great events; great revolutions of kingdoms, new prophets, destructive plagues, and other occurrences which bring distress and horror on mankind. No medical author of the fourteenth and fifteenth centuries omits an opportunity of representing them as among the general prognostics of great plagues; nor can we, for our part, regard the astrology of the Middle Ages as a mere offspring of superstition. It has not only, in common with all ideas which inspire and guide mankind, a high historical importance, entirely independent of its error or truth – for the influence of both is equally powerful – but there are also contained in it, as in alchemy, grand thoughts of antiquity, of which modern natural philosophy is so little ashamed that she claims them as her property. Foremost among these is the idea of general life which diffuses itself throughout the whole universe, expressed by the greatest Greek sages, and transmitted to the Middle Ages, through the new Platonic natural philosophy. To this impression of an universal organism, the assumption of a reciprocal influence of terrestrial bodies could not be foreign, nor did this cease to correspond with a higher view of nature, until astrologers overstepped the limits of human knowledge with frivolous and mystical calculations.

Guy de Chauliac considers the influence of the conjunction, which was held to be all-potent, as the chief general cause of the Black Plague; and the diseased state of bodies, the corruption of the fluids, debility, obstruction, and so forth, as the especial subordinate causes. By these, according to his opinion, the quality of the air, and of the other elements, was so altered that they set poisonous fluids in motion towards the inward parts of the body, in the same manner as the magnet attracts iron; whence there arose in the commencement fever and the spitting of blood; afterwards, however, a deposition in the form on glandular swellings and inflammatory boils. Herein the notion of an epidemic constitution was set forth clearly, and conformably to the spirit of the age. Of contagion, Guy de Chauliac was completely convinced. He sought to protect himself against it by the usual means; and it was probably he who advised Pope Clement VI. to shut himself up while the plague lasted. The preservation of this Pope's life, however, was most beneficial to the city of Avignon, for he loaded the poor with judicious acts of kindness, took care to have proper attendants provided, and paid physicians himself to afford assistance wherever human aid could avail – an advantage which, perhaps, no other city enjoyed. Nor was the treatment of plague-patients in Avignon by any means objectionable; for, after the usual depletions by bleeding and aperients, where circumstances required them, they endeavoured to bring the buboes to suppuration; they made incisions into the inflammatory boils, or burned them with a red-hot iron, a practice which at all times proves salutary, and in the Black Plague saved many lives. In this city, the Jews, who lived in a state of the greatest filth, were most severely visited, as also the Spaniards, whom Chalin accuses of great intemperance.

Still more distinct notions on the causes of the plague were stated to his contemporaries in the fourteenth century by Galeazzo di Santa Sofia, a learned man, a native of Padua, who likewise treated plague-patients at Vienna, though in what year is undetermined. He distinguishes carefully PESTILENCE from EPIDEMY and ENDEMY. The common notion of the two first accords exactly with that of an epidemic constitution, for both consist, according to him, in an unknown change or corruption of the air; with this difference, that pestilence calls forth diseases of different kinds; epidemy, on the contrary, always the same disease. As an example of an epidemy, he adduces a cough (influenza) which was observed in all climates at the same time without perceptible cause; but he recognised the approach of a pestilence, independently of unusual natural phenomena, by the more frequent occurrence of various kinds of fever, to which the modern physicians would assign a nervous and putrid character. The endemy originates, according to him, only in local telluric changes – in deleterious influences which develop themselves in the earth and in the water, without a corruption of the air. These notions were variously jumbled together in his time, like everything which human understanding separates by too fine a line of limitation. The estimation of cosmical influences, however, in the epidemy and pestilence, is well worthy of commendation; and Santa Sofia, in this respect, not only agrees with the most intelligent persons of the fourteenth and fifteenth centuries, but he has also promulgated an opinion which must, even now, serve as a foundation for our scarcely commenced investigations into cosmical influences. Pestilence and epidemy consist not in alterations of the four primary qualities, but in a corruption of the air, powerful, though quite immaterial, and not cognoscible by the senses – (corruptio aeris non substantialis, sed qualitativa) in a disproportion of the imponderables in the atmosphere, as it would be expressed by the moderns. The causes of the pestilence and epidemy are, first of all, astral influences, especially on occasions of planetary conjunctions; then extensive putrefaction of animal and vegetable bodies, and terrestrial corruptions (corruptio in terra): to which also bad diet and want may contribute. Santa Sofia considers the putrefaction of locusts, that had perished in the sea and were again thrown up, combined with astral and terrestrial influences, as the cause of the pestilence in the eventful year of the "Great Mortality."

All the fevers which were called forth by the pestilence are, according to him, of the putrid kind; for they originate principally from putridity of the heart's blood, which inevitably follows the inhalation of infected air. The Oriental Plague is, sometimes, but by no means always occasioned by pestilence, which imparts to it a character (qualitas occulta) hostile to human nature. It originates frequently from other causes, among which this physician was aware that contagion was to be reckoned; and it deserves to be remarked that he held epidemic smallpox and measles to be infallible forerunners of the plague, as do the physicians and people of the East at the present day.

In the exposition of his therapeutical views of the plague, a clearness of intellect is again shown by Santa Sofia, which reflects credit on the age. It seemed to him to depend,

1. On an evacuation of putrid matters by purgatives and bleeding; yet he did not sanction the employment of

these means indiscriminately and without consideration; least of all where the condition of the blood was healthy. He also declared himself decidedly against bleeding ad deliquium (venae sectio eradicativa).

2. Strengthening of the heart and prevention of putrescence.

3. Appropriate regimen.

4. Improvement of the air.

5. Appropriate treatment of tumid glands and inflammatory boils, with emollient, or even stimulating poultices (mustard, lily-bulbs), as well as with red-hot gold and iron.

6. Attention to prominent symptoms.

The stores of the Arabian pharmacy, which he brought into action to meet all these indications, were indeed very considerable; it is to be observed, however, that, for the most part, gentle means were accumulated, which, in case of abuse, would do no harm: for the character of the Arabian system of medicine, whose principles were everywhere followed at this time, was mildness and caution. On this account, too, we cannot believe that a very prolix treatise by Marsigli di Santa Sofia, a contemporary relative of Galeazzo, on the prevention and treatment of plague, can have caused much harm, although perhaps, even in the fourteenth century, an agreeable latitude and confident assertions respecting things which no mortal has investigated, or which it is quite a matter of indifference to distinguish, were considered as proofs of a valuable practical talent.

The agreement of contemporary and later writers shows that the published views of the most celebrated physicians of the fourteenth century were those generally adopted. Among these, Chalin de Vinario is the most experienced. Though devoted to astrology still more than his distinguished contemporary, he acknowledges the great power of terrestrial influences, and expresses himself very sensibly on the indisputable doctrine of contagion, endeavouring thereby to apologise for many surgeons and physicians of his time who neglected their duty. He asserted boldly and with truth, "that all epidemic diseases might become contagious, and all fevers epidemic," which attentive observers of all subsequent ages have confirmed.

He delivered his sentiments on bloodletting with sagacity, as an experienced physician; yet he was unable, as may be imagined, to moderate the desire for bleeding shown by the ignorant monks. He was averse to draw blood from the veins of patients under fourteen years of age; but counteracted inflammatory excitement in them by cupping, and endeavoured to moderate the inflammation of the tumid glands by leeches. Most of those who were bled, died; he therefore reserved this remedy for the plethoric; especially for the papal courtiers and the hypocritical priests, whom he saw gratifying their sensual desires, and imitating Epicurus, whilst they pompously pretended to follow Christ. He recommended burning the boils with a red-hot iron only in the plague without fever, which occurred in single cases; and was always ready to correct those overhasty surgeons who, with fire and violent remedies, did irremediable injury to their patients. Michael Savonarola, professor in Ferrara (1462), reasoning on the susceptibility of the human frame to the influence of pestilential infection, as the cause of such various modifications of disease, expresses himself as a modern physician would on this point; and an adoption of the principle of contagion was the foundation of his definition of the plague. No less worthy of observation are the views of the celebrated Valescus of Taranta, who, during the final visitation of the Black Death, in 1382, practised as a physician at Montpellier, and handed down to posterity what has been repeated in innumerable treatises on plague, which were written during the fifteenth and sixteenth centuries.

Of all these notions and views regarding the plague, whose development we have represented, there are two especially, which are prominent in historical importance:-

1. The opinion of learned physicians, that the pestilence, or epidemic constitution, is the parent of various kinds of disease; that the plague sometimes, indeed, but by no means always, originates from it: that, to speak in the language of the moderns, the pestilence bears the same relation to contagion that a predisposing cause does to an occasional cause; and

2. the universal conviction of the contagious power of that disease. Contagion gradually attracted more notice: it was thought that in it the most powerful occasional cause might be avoided; the possibility of protecting whole cities by separation became gradually more evident; and so horrifying was the recollection of the eventful year of the "Great

> Mortality," that before the close of the fourteenth century, ere the ill effects of the Black Plague had ceased, nations endeavoured to guard against the return of this enemy by an earnest and effectual defence.

Contagion gradually attracted more notice: it was thought that in it the most powerful occasional cause might be avoided; the possibility of protecting whole cities by separation became gradually more evident; and so horrifying was the recollection of the eventful year of the "Great Mortality," that before the close of the fourteenth century, ere the ill effects of the Black Plague had ceased, nations endeavoured to guard against the return of this enemy by an earnest and effectual defence.

The first regulation which was issued for this purpose, originated with Viscount Bernabo, and is dated the 17th January, 1374. "Every plague-patient was to be taken out of the city into the fields, there to die or to recover." Those who attended upon a plague-patient, were to remain apart for ten days before they again associated with anybody. The priests were to examine the diseased, and point out to special commissioners the persons infected, under punishment of the confiscation of their goods and of being burned alive. Whoever imported the plague, the state condemned his goods to confiscation. Finally, none except those who were appointed for that purpose were to attend plague-patients, under penalty of death and confiscation.

These orders, in correspondence with the spirit of the fourteenth century, are sufficiently decided to indicate a recollection of the good effects of confinement, and of keeping at a distance those suspected of having plague. It was said that Milan itself, by a rigorous barricade of three houses in which the plague had broken out, maintained itself free from the "Great Mortality" for a considerable time; and examples of the preservation of individual families, by means of a strict separation, were certainly very frequent. That these orders must have caused universal affliction from their uncommon severity, as we know to have been especially the case in the city of Reggio, may be easily conceived; but Bernabo did not suffer himself to be deterred from his purpose by fear – on the contrary, when the plague returned in the year 1383, he forbade the admission of people from infected places into his territories on pain of death. We have now, it is true, no account how far he succeeded; yet it is to be supposed that he arrested the disease, for it had long lost the property of the Black Death, to spread abroad in the air the contagious matter which proceeded from the lungs, charged with putridity, and to taint the atmosphere of whole cities by the vast numbers of the sick. Now that it had resumed its milder form, so that it infected only by contact, it admitted being confined within individual dwellings, as easily as in modern times.

Bernabo's example was imitated; nor was there any century more appropriate for recommending to governments strong regulations against the plague that the fourteenth; for when it broke out in Italy, in the year 1399, and still demanded new victims, it was for the sixteenth time, without reckoning frequent visitations of measles and smallpox. In this same year, Viscount John, in milder terms than his predecessor, ordered that no stranger should be admitted from infected places, and that the city gates should be strictly guarded. Infected houses were to be ventilated for at least eight or ten days, and purified from noxious vapours by fires, and by fumigations with balsamic and aromatic substances. Straw, rags, and the like were to be burned; and the bedsteads which had been used, set out for four days in the rain or the sunshine, so that by means of the one or the other, the morbific vapour might be destroyed. No one was to venture to make use of clothes or beds out of infected dwellings unless they had been previously washed and dried either at the fire or in the sun. People were, likewise, to avoid, as long as possible, occupying houses which had been frequented by plague-patients.

We cannot precisely perceive in these an advance towards general regulations; and perhaps people were convinced of the insurmountable impediments which opposed the separation of open inland countries, where bodies of people connected together could not be brought, even by the most obdurate severity, to renounce the habit of profitable intercourse.

Doubtless it is nature which has done the most to banish the Oriental plague from western Europe, where the increasing cultivation of the earth, and the advancing order in civilised society, have prevented it from remaining domesticated, which it most probably was in the more ancient times.

In the fifteenth century, during which it broke out seventeen times in different places in Europe, it was of the more consequence to oppose a barrier to its entrance from Asia, Africa, and Greece (which had become Turkish); for it would have been difficult for it to maintain itself indigenously any longer.

Among the southern commercial states, however, which were called on to make the greatest exertions to this end, it was principally Venice, formerly so severely attacked by the Black Plague, that put the necessary restraint upon perilous profits of the merchant. Until towards the end of the fifteenth century, the very considerable intercourse with the East was free and unimpeded. Ships of commercial cities had often brought over the plague: nay, the former irruption of the "Great Mortality" itself had been occasioned by navigators. For, as in the latter end of autumn, 1347, four ships full of plague-patients returned from the Levant to Genoa, the disease spread itself there with astonishing rapidity. On this account, in the following year, the Genoese forbade the entrance of suspected ships into their port. These sailed to Pisa and other cities on the coast, where already nature had made such mighty preparations for the reception of the Black Plague, and what we have already described took place in consequence.

In the year 1485, when, among the cities of northern Italy, Milan especially felt the scourge of the plague, a special Council of Health, consisting of three nobles, was established at Venice, who probably tried everything in their power to prevent the entrance of this disease, and gradually called into activity all those regulations which have served in later times as a pattern for the other southern states of Europe. Their endeavours were, however, not crowned with complete success; on which account their powers were increased, in the year 1504, by granting them the right of life and death over those who violated the regulations. Bills of health were probably first introduced in the year 1527, during a fatal plague which visited Italy for five years (1525-30), and called forth redoubled caution.

The first lazarettos were established upon islands at some distance from the city, seemingly as early as the year 1485. Here all strangers coming from places where the existence of plague was suspected were detained. If it appeared in the city itself, the sick were despatched with their families to what was called the Old Lazaretto, were there furnished with provisions and medicines, and when they were cured, were detained, together with all those who had had intercourse with them, still forty days longer in the New Lazaretto, situated on another island. All these regulations were every year improved, and their needful rigour was increased, so that from the year 1585 onwards, no appeal was allowed from the sentence of the Council of Health; and the other commercial nations gradually came to the support of the Venetians, by adopting corresponding regulations. Bills of health, however, were not general until the year 1665.

The appointment of a forty days' detention, whence quarantines derive their name, was not dictated by caprice, but probably had a medical origin, which is derivable in part from the doctrine of critical days; for the fortieth day, according to the most ancient notions, has been always regarded as the last of ardent diseases, and the limit of separation between these and those which are chronic. It was the custom to subject lying-in women for forty days to a more exact superintendence. There was a good deal also said in medical works of forty-day epochs in the formation of the foetus, not to mention that the alchemists expected more durable revolutions in forty days, which period they called the philosophical month.

This period being generally held to prevail in natural processes, it appeared reasonable to assume, and legally to establish it, as that required for the development of latent principles of contagion, since public regulations cannot dispense with decisions of this kind, even though they should not be wholly justified by the nature of the case. Great stress has likewise been laid on theological and legal grounds, which were certainly of greater weight in the fifteenth century than in the modern times.

On this matter, however, we cannot decide, since our only object here is to point out the origin of a political means of protection against a disease which has been the greatest impediment to civilisation within the memory of man; a means that, like Jenner's vaccine, after the smallpox had ravaged Europe for twelve hundred years, has diminished the check which mortality puts on the progress of civilisation, and thus given to the life and manners of the nations of this part of the world a new direction, the result of which we cannot foretell.

THE DANCING MANIA

CHAPTER I

THE DANCING MANIA IN GERMANY AND THE NETHERLANDS

***

SECT. 1 – ST. JOHN'S DANCE

The effects of the Black Death had not yet subsided, and the graves of millions of its victims were scarcely closed, when a strange delusion arose in Germany, which took possession of the minds of men, and, in spite of the divinity of our nature, hurried away body and soul into the magic circle of hellish superstition. It was a convulsion which in the most extraordinary manner infuriated the human frame, and excited the astonishment of contemporaries for more than two centuries, since which time it has never reappeared. It was called the dance of St. John or of St. Vitus, on account of the Bacchantic leaps by which it was characterised, and which gave to those affected, whilst performing their wild dance, and screaming and foaming with fury, all the appearance of persons possessed. It did not remain confined to particular localities, but was propagated by the sight of the sufferers, like a demoniacal epidemic, over the whole of Germany and the neighbouring countries to the northwest, which were already prepared for its reception by the prevailing opinions of the time.

So early as the year 1374, assemblages of men and women were seen at Aix-la-Chapelle, who had come out of Germany, and who, united by one common delusion, exhibited to the public both in the streets and in the churches the following strange spectacle. They formed circles hand in hand, and appearing to have lost all control over their senses, continued dancing, regardless of the bystanders, for hours together, in wild delirium, until at length they fell to the ground in a state of exhaustion. They then complained of extreme oppression, and groaned as if in the agonies of death, until they were swathed in cloths bound tightly round their waists, upon which they again recovered, and remained free from complaint until the next attack. This practice of swathing was resorted to on account of the tympany which followed these spasmodic ravings, but the bystanders frequently relieved patients in a less artificial manner, by thumping and trampling upon the parts affected. While dancing they neither saw nor heard, being insensible to external impressions through the senses, but were haunted by visions, their fancies conjuring up spirits whose names they shrieked out; and some of them afterwards asserted that they felt as if they had been immersed in a stream of blood, which obliged them to leap so high. Others, during the paroxysm, saw the heavens open and the Saviour enthroned with the Virgin Mary, according as the religious notions of the age were strangely and variously reflected in their imaginations.

Where the disease was completely developed, the attack commenced with epileptic convulsions. Those affected fell to the ground senseless, panting and labouring for breath. They foamed at the mouth, and suddenly springing up began their dance amidst strange contortions. Yet the malady doubtless made its appearance very variously, and was modified by temporary or local circumstances, whereof non-medical contemporaries but imperfectly noted the essential particulars, accustomed as they were to confound their observation of natural events with their notions of the world of spirits.

It was but a few months ere this demoniacal disease had spread from Aix-la-Chapelle, where it appeared in July, over the neighbouring Netherlands. In Liege, Utrecht, Tongres, and many other towns of Belgium, the dancers appeared with garlands in their hair, and their waists girt with cloths, that they might, as soon as the paroxysm was over, receive immediate relief on the attack of the tympany. This bandage was, by the insertion of a stick, easily twisted tight: many, however, obtained more relief from kicks and blows, which they found numbers of persons ready to administer: for, wherever the dancers appeared, the people assembled in crowds to gratify their curiosity with the frightful spectacle. At length the increasing number of the affected excited no less anxiety than the attention that was paid to them. In towns and villages they took possession of the religious houses, processions were everywhere instituted on their account, and masses were said and hymns were sung, while the disease itself, of the demoniacal origin of which no one entertained the least doubt, excited everywhere astonishment and horror. In Liege the priests had recourse to exorcisms, and endeavoured by every means in their power to allay an evil which threatened so much danger to themselves; for the possessed assembling in multitudes, frequently poured forth imprecations against them, and menaced their destruction. They intimidated the people also to such a degree that there was an express ordinance issued that no one should make any but square-toed shoes, because these fanatics had manifested a morbid dislike to the pointed shoes which had come into fashion immediately after the "Great Mortality" in 1350. They were still more irritated at the sight of red colours, the influence of which on the disordered nerves might lead us to imagine an extraordinary accordance between this spasmodic malady and the condition of infuriated animals; but in the St. John's dancers this excitement was probably connected with apparitions consequent upon their convulsions. There were likewise some of them who were unable to endure the sight of persons weeping. The clergy seemed to become daily more and more confirmed in their belief that those who were affected were a kind of sectarians, and on this account they hastened their exorcisms as much as possible, in order that the evil might not spread amongst the higher classes, for hitherto scarcely any but the poor had been attacked, and the few people of respectability among the laity and clergy who were to be found among them, were persons whose natural frivolity was unable to withstand the excitement of novelty, even though it proceeded from a demoniacal influence. Some of the affected had indeed themselves declared, when under the influence of priestly forms of exorcism, that if the demons had been allowed only a few weeks' more time, they would have entered the bodies of the nobility and princes, and through these have destroyed the clergy. Assertions of this sort, which those possessed uttered whilst in a state which may be compared with that of magnetic sleep, obtained general belief, and passed from mouth to mouth with wonderful additions. The priesthood were, on this account, so much the more zealous in their endeavours to anticipate every dangerous excitement of the people, as if the existing order of things could have been seriously threatened by such incoherent ravings. Their exertions were effectual, for exorcism was a powerful remedy in the fourteenth century; or it might perhaps be that this wild infatuation terminated in consequence of the exhaustion which naturally ensued from it; at all events, in the course of ten or eleven months the St. John's dancers were no longer to be found in any of the cities of Belgium. The evil, however, was too deeply rooted to give way altogether to such feeble attacks.

A few months after this dancing malady had made its appearance at Aix-la-Chapelle, it broke out at Cologne, where the number of those possessed amounted to more than five hundred, and about the same time at Metz, the streets of which place are said to have been filled with eleven hundred dancers. Peasants left their ploughs, mechanics their workshops, housewives their domestic duties, to join the wild revels, and this rich commercial city became the scene of the most ruinous disorder. Secret desires were excited, and but too often found opportunities for wild enjoyment; and numerous beggars, stimulated by vice and misery, availed themselves of this new complaint to gain a temporary livelihood. Girls and boys quitted their parents, and servants their masters, to amuse themselves at the dances of those possessed, and greedily imbibed the poison of mental infection. Above a hundred unmarried women were seen raving about in consecrated and unconsecrated places, and the consequences were soon perceived. Gangs of idle vagabonds, who understood how to imitate to the life the gestures and convulsions of those really affected, roved from place to place seeking maintenance and adventures, and thus, wherever they went, spreading this disgusting spasmodic disease like a plague; for in maladies of this kind the susceptible are infected as easily by the appearance as by the reality. At last it was found necessary to drive away these mischievous guests, who were equally inaccessible to the exorcisms of the priests and the remedies of the physicians. It was not, however, until after four months that the Rhenish cities were able to suppress these impostures, which had so alarmingly increased the original evil. In the meantime, when once called into existence, the plague crept on, and found abundant food in the tone of thought which prevailed in the fourteenth and fifteenth centuries, and even, though in a minor degree, throughout the sixteenth and seventeenth, causing a permanent disorder of the mind, and exhibiting in those cities to whose inhabitants it was a novelty, scenes as strange as they were detestable.

SECT. 2 – ST. VITUS'S DANCE

Strasburg was visited by the "Dancing Plague" in the year 1418, and the same infatuation existed among the people there, as in the towns of Belgium and the Lower Rhine. Many who were seized at the sight of those affected, excited attention at first by their confused and absurd behaviour, and then by their constantly following swarms of dancers. These were seen day and night passing through the streets, accompanied by musicians playing on bagpipes, and by innumerable spectators attracted by curiosity, to which were added anxious parents and relations, who came to look after those among the misguided multitude who belonged to their respective families. Imposture and profligacy played their part in this city also, but the morbid delusion itself seems to have predominated. On this account religion could only bring provisional aid, and therefore the town council benevolently took an interest in the afflicted. They divided them into separate parties, to each of which they appointed responsible superintendents to protect them from harm, and perhaps also to restrain their turbulence. They were thus conducted on foot and in carriages to the chapels of St. Vitus, near Zabern and Rotestein, where priests were in attendance to work upon their misguided minds by masses and other religious ceremonies. After divine worship was completed, they were led in solemn procession to the altar, where they made some small offering of alms, and where it is probable that many were, through the influence of devotion and the sanctity of the place, cured of this lamentable aberration. It is worthy of observation, at all events, that the Dancing Mania did not recommence at the altars of the saint, and that from him alone assistance was implored, and through his miraculous interposition a cure was expected, which was beyond the reach of human skill. The personal history of St. Vitus is by no means important in this matter. He was a Sicilian youth, who, together with Modestus and Crescentia, suffered martyrdom at the time of the persecution of the Christians, under Diocletian, in the year 303. The legends respecting him are obscure, and he would certainly have been passed over without notice among the innumerable apocryphal martyrs of the first centuries, had not the transfer of his body to St. Denys, and thence, in the year 836, to Corvey, raised him to a higher rank. From this time forth it may be supposed that many miracles were manifested at his new sepulchre, which were of essential service in confirming the Roman faith among the Germans, and St. Vitus was soon ranked among the fourteen saintly helpers (Nothhelfer or Apotheker). His altars were multiplied, and the people had recourse to them in all kinds of distresses, and revered him as a powerful intercessor. As the worship of these saints was, however, at that time stripped of all historical connections, which were purposely obliterated by the priesthood, a legend was invented at the beginning of the fifteenth century, or perhaps even so early as the fourteenth, that St. Vitus had, just before he bent his neck to the sword, prayed to God that he might protect from the Dancing Mania all those who should solemnise the day of his commemoration, and fast upon its eve, and that thereupon a voice from heaven was heard, saying, "Vitus, thy prayer is accepted." Thus St. Vitus became the patron saint of those afflicted with the Dancing Plague, as St. Martin of Tours was at one time the succourer of persons in smallpox, St. Antonius of those suffering under the "hellish fire," and as St. Margaret was the Juno Lucina of puerperal women.

SECT. 3 – CAUSES

The connection which John the Baptist had with the Dancing Mania of the fourteenth century was of a totally different character. He was originally far from being a protecting saint to those who were attacked, or one who would be likely to give them relief from a malady considered as the work of the devil. On the contrary, the manner in which he was worshipped afforded an important and very evident cause for its development. From the remotest period, perhaps even so far back as the fourth century, St. John's day was solemnised with all sorts of strange and rude customs, of which the originally mystical meaning was variously disfigured among different nations by superadded relics of heathenism. Thus the Germans transferred to the festival of St. John's day an ancient heathen usage, the kindling of the "Nodfyr," which was forbidden them by St. Boniface, and the belief subsists even to the present day that people and animals that have leaped through these flames, or their smoke, are protected for a whole year from fevers and other diseases, as if by a kind of baptism by fire. Bacchanalian dances, which have originated in similar causes among all the rude nations of the earth, and the wild extravagancies of a heated imagination, were the constant accompaniments of this half-heathen, half-Christian festival. At the period of which we are treating, however, the Germans were not the only people who gave way to the ebullitions of fanaticism in keeping the festival of St. John the Baptist. Similar customs were also to be found among the nations of Southern Europe and of Asia, and it is more than probable that the Greeks transferred to the festival of John the Baptist, who is also held in high esteem among the Mahomedans, a part of their Bacchanalian mysteries, an absurdity of a kind which is but too frequently met with in human affairs. How far a remembrance of the history of St. John's death may have had an influence on this occasion, we would leave learned theologians to decide. It is only of importance here to add that in Abyssinia, a country entirely separated from Europe, where Christianity has maintained itself in its primeval simplicity against Mahomedanism, John is to this day worshipped, as protecting saint of those who are attacked with the dancing malady. In these fragments of the dominion of mysticism and superstition, historical connection is not to be found.

When we observe, however, that the first dancers in Aix-la-Chapelle appeared in July with St. John's name in their mouths, the conjecture is probable that the wild revels of St. John's day, A.D. 1374, gave rise to this mental plague, which thenceforth has visited so many thousands with incurable aberration of mind, and disgusting distortions of body.

This is rendered so much the more probable because some months previously the districts in the neighbourhood of the Rhine and the Main had met with great disasters. So early as February, both these rivers had overflowed their banks to a great extent; the walls of the town of Cologne, on the side next the Rhine, had fallen down, and a great many villages had been reduced to the utmost distress. To this was added the miserable condition of western and southern Germany. Neither law nor edict could suppress the incessant feuds of the Barons, and in Franconia especially, the ancient times of club law appeared to be revived. Security of property there was none; arbitrary will everywhere prevailed; corruption of morals and rude power rarely met with even a feeble opposition; whence it arose that the cruel, but lucrative, persecutions of the Jews were in many places still practised through the whole of this century with their wonted ferocity. Thus, throughout the western parts of Germany, and especially in the districts bordering on the Rhine, there was a wretched and oppressed populace; and if we take into consideration that among their numerous bands many wandered about, whose consciences were tormented with the recollection of the crimes which they had committed during the prevalence of the Black Plague, we shall comprehend how their despair sought relief in the intoxication of an artificial delirium. There is hence good ground for supposing that the frantic celebration of the festival of St. John, A.D. 1374, only served to bring to a crisis a malady which had been long impending; and if we would further inquire how a hitherto harmless usage, which like many others had but served to keep up superstition, could degenerate into so serious a disease, we must take into account the unusual excitement of men's minds, and the consequences of wretchedness and want. The bowels, which in many were debilitated by hunger and bad food, were precisely the parts which in most cases were attacked with excruciating pain, and the tympanitic state of the intestines points out to the intelligent physician an origin of the disorder which is well worth consideration.

SECT. 4 – MORE ANCIENT DANCING PLAGUES

The Dancing Mania of the year 1374 was, in fact, no new disease, but a phenomenon well known in the Middle Ages, of which many wondrous stories were traditionally current among the people. In the year 1237 upwards of a hundred children were said to have been suddenly seized with this disease at Erfurt, and to have proceeded dancing and jumping along the road to Arnstadt. When they arrived at that place they fell exhausted to the ground, and, according to an account of an old chronicle, many of them, after they were taken home by their parents, died, and the rest remained affected, to the end of their lives, with a permanent tremor. Another occurrence was related to have taken place on the Moselle Bridge at Utrecht, on the 17th day of June, A.D. 1278, when two hundred fanatics began to dance, and would not desist until a priest passed, who was carrying the Host to a person that was sick, upon which, as if in punishment of their crime, the bridge gave way, and they were all drowned. A similar event also occurred so early as the year 1027, near the convent church of Kolbig, not far from Bernburg. According to an oft-repeated tradition, eighteen peasants, some of whose names are still preserved, are said to have disturbed divine service on Christmas Eve by dancing and brawling in the churchyard, whereupon the priest, Ruprecht, inflicted a curse upon them, that they should dance and scream for a whole year without ceasing. This curse is stated to have been completely fulfilled, so that the unfortunate sufferers at length sank knee-deep into the earth, and remained the whole time without nourishment, until they were finally released by the intercession of two pious bishops. It is said that, upon this, they fell into a deep sleep, which lasted three days, and that four of them died; the rest continuing to suffer all their lives from a trembling of their limbs. It is not worth while to separate what may have been true, and what the addition of crafty priests, in this strangely distorted story. It is sufficient that it was believed, and related with astonishment and horror, throughout the Middle Ages; so that when there was any exciting cause for this delirious raving and wild rage for dancing, it failed not to produce its effects upon men whose thoughts were given up to a belief in wonders and apparitions.

This disposition of mind, altogether so peculiar to the Middle Ages, and which, happily for mankind, has yielded to an improved state of civilisation and the diffusion of popular instruction, accounts for the origin and long duration of this extraordinary mental disorder. The good sense of the people recoiled with horror and aversion from this heavy plague, which, whenever malevolent persons wished to curse their bitterest enemies and adversaries, was long after used as a malediction. The indignation also that was felt by the people at large against the immorality of the age, was proved by their ascribing this frightful affliction to the inefficacy of baptism by unchaste priests, as if innocent children were doomed to atone, in after years, for this desecration of the sacrament administered by unholy hands. We have already mentioned what perils the priests in the Netherlands incurred from this belief.

They now, indeed, endeavoured to hasten their reconciliation with the irritated, and, at that time, very degenerate people, by exorcisms, which, with some, procured them greater respect than ever, because they thus visibly restored thousands of those who were affected. In general, however, there prevailed a want of confidence in their efficacy, and then the sacred rites had as little power in arresting the progress of this deeply-rooted malady as the prayers and holy services subsequently had at the altars of the greatly-revered martyr St. Vitus. We may therefore ascribe it to accident merely, and to a certain aversion to this demoniacal disease, which seemed to lie beyond the reach of human skill, that we meet with but few and imperfect notices of the St. Vitus's dance in the second half of the fifteenth century. The highly-coloured descriptions of the sixteenth century contradict the notion that this mental plague had in any degree diminished in its severity, and not a single fact is to be found which supports the opinion that any one of the essential symptoms of the disease, not even excepting the tympany, had disappeared, or that the disorder itself had become milder in its attacks. The physicians never, as it seems, throughout the whole of the fifteenth century, undertook the treatment of the Dancing Mania, which, according to the prevailing notions, appertained exclusively to the servants of the Church. Against demoniacal disorders they had no remedies, and though some at first did promulgate the opinion that the malady had its origin in natural circumstances, such as a hot temperament, and other causes named in the phraseology of the schools, yet these opinions were the less examined as it did not appear worth while to divide with a jealous priesthood the care of a host of fanatical vagabonds and beggars.

SECT. 5 – PHYSICIANS

It was not until the beginning of the sixteenth century that the St. Vitus's dance was made the subject of medical research, and stripped of its unhallowed character as a work of demons. This was effected by Paracelsus, that mighty but, as yet, scarcely comprehended reformer of medicine, whose aim it was to withdraw diseases from the pale of miraculous interpositions and saintly influences, and explain their causes upon principles deduced from his knowledge of the human frame. "We will not, however, admit that the saints have power to inflict diseases, and that these ought to be named after them, although many there are who, in their theology, lay great stress on this supposition, ascribing them rather to God than to nature, which is but idle talk. We dislike such nonsensical gossip as is not supported by symptoms, but only by faith – a thing which is not human, whereon the gods themselves set no value."

Such were the words which Paracelsus addressed to his contemporaries, who were, as yet, incapable of appreciating doctrines of this sort; for the belief in enchantment still remained everywhere unshaken, and faith in the world of spirits still held men's minds in so close a bondage that thousands were, according to their own conviction, given up as a prey to the devil; while at the command of religion, as well as of law, countless piles were lighted, by the flames of which human society was to be purified.

Paracelsus divides the St. Vitus's dance into three kinds. First, that which arises from imagination (Vitista, Chorea imaginativa, aestimativa), by which the original Dancing Plague is to be understood. Secondly, that which arises from sensual desires, depending on the will (Chorea lasciva). Thirdly, that which arises from corporeal causes (Chorea naturalis, coacta), which, according to a strange notion of his own, he explained by maintaining that in certain vessels which are susceptible of an internal pruriency, and thence produce laughter, the blood is set in commotion in consequence of an alteration in the vital spirits, whereby involuntary fits of intoxicating joy and a propensity to dance are occasioned. To this notion he was, no doubt, led from having observed a milder form of St. Vitus's dance, not uncommon in his time, which was accompanied by involuntary laughter; and which bore a resemblance to the hysterical laughter of the moderns, except that it was characterised by more pleasurable sensations and by an extravagant propensity to dance. There was no howling, screaming, and jumping, as in the severer form; neither was the disposition to dance by any means insuperable. Patients thus affected, although they had not a complete control over their understandings, yet were sufficiently self-possessed during the attack to obey the directions which they received. There were even some among them who did not dance at all, but only felt an involuntary impulse to allay the internal sense of disquietude, which is the usual forerunner of an attack of this kind, by laughter and quick walking carried to the extent of producing fatigue. This disorder, so different from the original type, evidently approximates to the modern chorea; or, rather, is in perfect accordance with it, even to the less essential symptom of laughter. A mitigation in the form of the Dancing Mania had thus clearly taken place at the commencement of the sixteenth century.

On the communication of the St. Vitus's dance by sympathy, Paracelsus, in his peculiar language, expresses himself with great spirit, and shows a profound knowledge of the nature of sensual impressions, which find their way to the heart – the seat of joys and emotions – which overpower the opposition of reason; and whilst "all other qualities and natures" are subdued, incessantly impel the patient, in consequence of his original compliance, and his all-conquering imagination, to imitate what he has seen. On his treatment of the disease we cannot bestow any great praise, but must be content with the remark that it was in conformity with the notions of the age in which he lived. For the first kind, which often originated in passionate excitement, he had a mental remedy, the efficacy of which is not to be despised, if we estimate its value in connection with the prevalent opinions of those times. The patient was to make an image of himself in wax or resin, and by an effort of thought to concentrate all his blasphemies and sins in it. "Without the intervention of any other persons, to set his whole mind and thoughts concerning these oaths in the image;" and when he had succeeded in this, he was to burn the image, so that not a particle of it should remain. In all this there was no mention made of St. Vitus, or any of the other mediatory saints, which is accounted for by the circumstance that at this time an open rebellion against the Romish Church had begun, and the worship of saints was by many rejected as idolatrous. For the second kind of St. Vitus's dance, arising from sensual irritation, with which women were far more frequently affected than men, Paracelsus recommended harsh treatment and strict fasting. He directed that the patients should be deprived of their liberty; placed in solitary confinement, and made to sit in an uncomfortable place, until their misery brought them to their senses and to a feeling of penitence. He then permitted them gradually to return to their accustomed habits. Severe corporal chastisement was not omitted; but, on the other hand, angry resistance on the part of the patient was to be sedulously avoided, on the ground that it might increase his malady, or even destroy him: moreover, where it seemed proper, Paracelsus allayed the excitement of the nerves by immersion in cold water. On the treatment of the third kind we shall not here enlarge. It was to be effected by all sorts of wonderful remedies, composed of the quintessences; and it would require, to render it intelligible, a more extended exposition of peculiar principles than suits our present purpose.

SECT. 6 – DECLINE AND TERMINATION OF THE DANCING PLAGUE

About this time the St. Vitus's dance began to decline, so that milder forms of it appeared more frequently, while the severer cases became more rare;

and even in these, some of the important symptoms gradually disappeared. Paracelsus makes no mention of the tympanites as taking place after the attacks, although it may occasionally have occurred; and Schenck von Graffenberg, a celebrated physician of the latter half of the sixteenth century, speaks of this disease as having been frequent only in the time of his forefathers; his descriptions, however, are applicable to the whole of that century, and to the close of the fifteenth. The St. Vitus's dance attacked people of all stations, especially those who led a sedentary life, such as shoemakers and tailors; but even the most robust peasants abandoned their labours in the fields, as if they were possessed by evil spirits; and thus those affected were seen assembling indiscriminately, from time to time, at certain appointed places, and, unless prevented by the lookers-on, continuing to dance without intermission, until their very last breath was expended. Their fury and extravagance of demeanour so completely deprived them of their senses, that many of them dashed their brains out against the walls and corners of buildings, or rushed headlong into rapid rivers, where they found a watery grave. Roaring and foaming as they were, the bystanders could only succeed in restraining them by placing benches and chairs in their way, so that, by the high leaps they were thus tempted to take, their strength might be exhausted. As soon as this was the case, they fell as it were lifeless to the ground, and, by very slow degrees, again recovered their strength. Many there were who, even with all this exertion, had not expended the violence of the tempest which raged within them, but awoke with newly-revived powers, and again and again mixed with the crowd of dancers, until at length the violent excitement of their disordered nerves was allayed by the great involuntary exertion of their limbs; and the mental disorder was calmed by the extreme exhaustion of the body. Thus the attacks themselves were in these cases, as in their nature they are in all nervous complaints, necessary crises of an inward morbid condition which was transferred from the sensorium to the nerves of motion, and, at an earlier period, to the abdominal plexus, where a deep-seated derangement of the system was perceptible from the secretion of flatus in the intestines.

The cure effected by these stormy attacks was in many cases so perfect, that some patients returned to the factory or the plough as if nothing had happened. Others, on the contrary, paid the penalty of their folly by so total a loss of power, that they could not regain their former health, even by the employment of the most strengthening remedies. Medical men were astonished to observe that women in an advanced state of pregnancy were capable of going through an attack of the disease without the slightest injury to their offspring, which they protected merely by a bandage passed round the waist. Cases of this kind were not infrequent so late as Schenck's time. That patients should be violently affected by music, and their paroxysms brought on and increased by it, is natural with such nervous disorders, where deeper impressions are made through the ear, which is the most intellectual of all the organs, than through any of the other senses. On this account the magistrates hired musicians for the purpose of carrying the St. Vitus's dancers so much the quicker through the attacks, and directed that athletic men should be sent among them in order to complete the exhaustion, which had been often observed to produce a good effect. At the same time there was a prohibition against wearing red garments, because, at the sight of this colour, those affected became so furious that they flew at the persons who wore it, and were so bent upon doing them an injury that they could with difficulty be restrained. They frequently tore their own clothes whilst in the paroxysm, and were guilty of other improprieties, so that the more opulent employed confidential attendants to accompany them, and to take care that they did no harm either to themselves or others. This extraordinary disease was, however, so greatly mitigated in Schenck's time, that the St. Vitus's dancers had long since ceased to stroll from town to town; and that physician, like Paracelsus, makes no mention of the tympanitic inflation of the bowels. Moreover, most of those affected were only annually visited by attacks; and the occasion of them was so manifestly referable to the prevailing notions of that period, that if the unqualified belief in the supernatural agency of saints could have been abolished, they would not have had any return of the complaint. Throughout the whole of June, prior to the festival of St. John, patients felt a disquietude and restlessness which they were unable to overcome. They were dejected, timid, and anxious; wandered about in an unsettled state, being tormented with twitching pains, which seized them suddenly in different parts, and eagerly expected the eve of St. Johns day, in the confident hope that by dancing at the altars of this saint, or of St. Vitus (for in the Breisgau aid was equally sought from both), they would be freed from all their sufferings. This hope was not disappointed; and they remained, for the rest of the year, exempt from any further attack, after having thus, by dancing and raving for three hours, satisfied an irresistible demand of nature. There were at that period two chapels in the Breisgau visited by the St. Vitus's dancers; namely, the Chapel of St. Vitus at Biessen, near Breisach, and that of St. John, near Wasenweiler; and it is probable that in the southwest of Germany the disease was still in existence in the seventeenth century.

However, it grew every year more rare, so that at the beginning of the seventeenth century it was observed only occasionally in its ancient form. Thus in the spring of the year 1623, G. Horst saw some women who annually performed a pilgrimage to St. Vitus's chapel at Drefelhausen, near Weissenstein, in the territory of Ulm, that they might wait for their dancing fit there, in the same manner as those in the Breisgau did, according to Schenck's account. They were not satisfied, however, with a dance of three hours' duration, but continued day and night in a state of mental aberration, like persons in an ecstasy, until they fell exhausted to the ground; and when they came to themselves again they felt relieved from a distressing uneasiness and painful sensation of weight in their bodies, of which they had complained for several weeks prior to St. Vitus's Day.

After this commotion they remained well for the whole year; and such was their faith in the protecting power of the saint, that one of them had visited this shrine at Drefelhausen more than twenty times, and another had already kept the saint's day for the thirty-second time at this sacred station.

The dancing fit itself was excited here, as it probably was in other places, by music, from the effects of which the patients were thrown into a state of convulsion. Many concurrent testimonies serve to show that music generally contributed much to the continuance of the St. Vitus's dance, originated and increased its paroxysms, and was sometimes the cause of their mitigation. So early as the fourteenth century the swarms of St. John's dancers were accompanied by minstrels playing upon noisy instruments, who roused their morbid feelings; and it may readily be supposed that by the performance of lively melodies, and the stimulating effects which the shrill tones of fifes and trumpets would produce, a paroxysm that was perhaps but slight in itself, might, in many cases, be increased to the most outrageous fury, such as in later times was purposely induced in order that the force of the disease might be exhausted by the violence of its attack. Moreover, by means of intoxicating music a kind of demoniacal festival for the rude multitude was established, which had the effect of spreading this unhappy malady wider and wider. Soft harmony was, however, employed to calm the excitement of those affected, and it is mentioned as a character of the tunes played with this view to the St. Vitus's dancers, that they contained transitions from a quick to a slow measure, and passed gradually from a high to a low key. It is to be regretted that no trace of this music has reached out times, which is owing partly to the disastrous events of the seventeenth century, and partly to the circumstance that the disorder was looked upon as entirely national, and only incidentally considered worthy of notice by foreign men of learning. If the St. Vitus's dance was already on the decline at the commencement of the seventeenth century, the subsequent events were altogether adverse to its continuance. Wars carried on with animosity, and with various success, for thirty years, shook the west of Europe; and although the unspeakable calamities which they brought upon Germany, both during their continuance and in their immediate consequences, were by no means favourable to the advance of knowledge, yet, with the vehemence of a purifying fire, they gradually effected the intellectual regeneration of the Germans; superstition, in her ancient form, never again appeared, and the belief in the dominion of spirits, which prevailed in the middle ages, lost for ever its once formidable power.

CHAPTER II

THE DANCING MANIA IN ITALY

***

SECT. 1 – TARANTISM

It was of the utmost advantage to the St. Vitus's dancers that they made choice of a favourite patron saint; for, not to mention that people were inclined to compare them to the possessed with evil spirits described in the Bible, and thence to consider them as innocent victims to the power of Satan, the name of their great intercessor recommended them to general commiseration, and a magic boundary was thus set to every harsh feeling, which might otherwise have proved hostile to their safety. Other fanatics were not so fortunate, being often treated with the most relentless cruelty, whenever the notions of the middle ages either excused or commanded it as a religious duty. Thus, passing over the innumerable instances of the burning of witches, who were, after all, only labouring under a delusion, the Teutonic knights in Prussia not unfrequently condemned those maniacs to the stake who imagined themselves to be metamorphosed into wolves – an extraordinary species of insanity, which, having existed in Greece before our era, spread, in process of time over Europe, so that it was communicated not only to the Romaic, but also to the German and Sarmatian nations, and descended from the ancients as a legacy of affliction to posterity. In modern times Lycanthropy – such was the name given to this infatuation – has vanished from the earth, but it is nevertheless well worthy the consideration of the observer of human aberrations, and a history of it by some writer who is equally well acquainted with the middle ages as with antiquity is still a desideratum. We leave it for the present without further notice, and turn to a malady most extraordinary in all its phenomena, having a close connection with the St. Vitus's dance, and, by a comparison of facts which are altogether similar, affording us an instructive subject for contemplation. We allude to the disease called Tarantism, which made its first appearance in Apulia, and thence spread over the other provinces of Italy, where, during some centuries, it prevailed as a great epidemic. In the present times, it has vanished, or at least has lost altogether its original importance, like the St. Vitus's dance, lycanthropy, and witchcraft.

SECT. 2 – MOST ANCIENT TRACES – CAUSES

The learned Nicholas Perotti gives the earliest account of this strange disorder. Nobody had the least doubt that it was caused by the bite of the tarantula, a ground-spider common in Apulia: and the fear of this insect was so general that its bite was in all probability much oftener imagined, or the sting of some other kind of insect mistaken for it, than actually received. The word tarantula is apparently the same as terrantola, a name given by the Italians to the stellio of the old Romans, which was a kind of lizard, said to be poisonous, and invested by credulity with such extraordinary qualities, that, like the serpent of the Mosaic account of the Creation, it personified, in the imaginations of the vulgar, the notion of cunning, so that even the jurists designated a cunning fraud by the appellation of a "stellionatus." Perotti expressly assures us that this reptile was called by the Romans tarantula; and since he himself, who was one of the most distinguished authors of his time, strangely confounds spiders and lizards together, so that he considers the Apulian tarantula, which he ranks among the class of spiders, to have the same meaning as the kind of lizard called [Greek text], it is the less extraordinary that the unlearned country people of Apulia should confound the much-dreaded ground-spider with the fabulous star-lizard, and appropriate to the one the name of the other. The derivation of the word tarantula, from the city of Tarentum, or the river Thara, in Apulia, on the banks of which this insect is said to have been most frequently found, or, at least, its bite to have had the most venomous effect, seems not to be supported by authority. So much for the name of this famous spider, which, unless we are greatly mistaken, throws no light whatever upon the nature of the disease in question. Naturalists who, possessing a knowledge of the past, should not misapply their talents by employing them in establishing the dry distinction of forms, would find here much that calls for research, and their efforts would clear up many a perplexing obscurity.

Perotti states that the tarantula - that is, the spider so called - was not met with in Italy in former times, but that in his day it had become common, especially in Apulia, as well as in some other districts. He deserves, however, no great confidence as a naturalist, notwithstanding his having delivered lectures in Bologna on medicine and other sciences. He at least has neglected to prove his assertion, which is not borne out by any analogous phenomenon observed in modern times with regard to the history of the spider species. It is by no means to be admitted that the tarantula did not make its appearance in Italy before the disease ascribed to its bite became remarkable, even though tempests more violent than those unexampled storms which arose at the time of the Black Death in the middle of the fourteenth century had set the insect world in motion; for the spider is little if at all susceptible of those cosmical influences which at times multiply locusts and other winged insects to a wonderful extent, and compel them to migrate.

The symptoms which Perotti enumerates as consequent on the bite of the tarantula agree very exactly with those described by later writers. Those who were bitten, generally fell into a state of melancholy, and appeared to be stupefied, and scarcely in possession of their senses. This condition was, in many cases, united with so great a sensibility to music, that at the very first tones of their favourite melodies they sprang up, shouting for joy, and danced on without intermission, until they sank to the ground exhausted and almost lifeless. In others, the disease did not take this cheerful turn. They wept constantly, and as if pining away with some unsatisfied desire, spent their days in the greatest misery and anxiety. Others, again, in morbid fits of love, cast their longing looks on women, and instances of death are recorded, which are said to have occurred under a paroxysm of either laughing or weeping.

From this description, incomplete as it is, we may easily gather that tarantism, the essential symptoms of which are mentioned in it, could not have originated in the fifteenth century, to which Perotti's account refers; for that author speaks of it as a well-known malady, and states that the omission to notice it by older writers was to be ascribed solely to the want of education in Apulia, the only province probably where the disease at that time prevailed. A nervous disorder that had arrived at so high a degree of development must have been long in existence, and doubtless had required an elaborate preparation by the concurrence of general causes.

The symptoms which followed the bite of venomous spiders were well known to the ancients, and had excited the attention of their best observers, who agree in their descriptions of them. It is probable that among the numerous species of their phalangium, the Apulian tarantula is included, but it is difficult to determine this point with certainty, more especially because in Italy the tarantula was not the only insect which caused this nervous affection, similar results being likewise attributed to the bite of the scorpion. Lividity of the whole body, as well as of the countenance, difficulty of speech, tremor of the limbs, icy coldness, pale urine, depression of spirits, headache, a flow of tears, nausea, vomiting, sexual excitement, flatulence, syncope, dysuria, watchfulness, lethargy, even death itself, were cited by them as the consequences of being bitten by venomous spiders, and they made little distinction as to their kinds. To these symptoms we may add the strange rumour, repeated throughout the middle ages, that persons who were bitten, ejected by the bowels and kidneys, and even by vomiting, substances resembling a spider's web.

Nowhere, however, do we find any mention made that those affected felt an irresistible propensity to dancing, or that they were accidentally cured by it. Even Constantine of Africa, who lived 500 years after Aetius, and, as the most learned physician of the school of Salerno, would certainly not have passed over so acceptable a subject of remark, knows nothing of such a memorable course of this disease arising from poison, and merely repeats the observations of his Greek predecessors. Gariopontus, a Salernian physician of the eleventh century, was the first to describe a kind of insanity, the remote affinity of which to the tarantula disease is rendered apparent by a very striking symptom. The patients in their sudden attacks behaved like maniacs, sprang up, throwing their arms about with wild movements, and, if perchance a sword was at hand, they wounded themselves and others, so that it became necessary carefully to secure them. They imagined that they heard voices and various kinds of sounds, and if, during this state of illusion, the tones of a favourite instrument happened to catch their ear, they commenced a spasmodic dance, or ran with the utmost energy which they could muster until they were totally exhausted. These dangerous maniacs, who, it would seem, appeared in considerable numbers, were looked upon as a legion of devils, but on the causes of their malady this obscure writer adds nothing further than that he believes (oddly enough) that it may sometimes be excited by the bite of a mad dog. He calls the disease Anteneasmus, by which is meant no doubt the Enthusiasmus of the Greek physicians. We cite this phenomenon as an important forerunner of tarantism, under the conviction that we have thus added to the evidence that the development of this latter must have been founded on circumstances which existed from the twelfth to the end of the fourteenth century; for the origin of tarantism itself is referable, with the utmost probability, to a period between the middle and the end of this century, and is consequently contemporaneous with that of the St. Vitus's dance (1374). The influence of the Roman Catholic religion, connected as this was, in the middle ages, with the pomp of processions, with public exercises of penance, and with innumerable practices which strongly excited the imaginations of its votaries, certainly brought the mind to a very favourable state for the reception of a nervous disorder. Accordingly, so long as the doctrines of Christianity were blended with so much mysticism, these unhallowed disorders prevailed to an important extent, and even in our own days we find them propagated with the greatest facility where the existence of superstition produces the same effect, in more limited districts, as it once did among whole nations. But this is not all. Every country in Europe, and Italy perhaps more than any other, was visited during the middle ages by frightful plagues, which followed each other in such quick succession that they gave the exhausted people scarcely any time for recovery. The Oriental bubo-plague ravaged Italy sixteen times between the years 1119 and 1340. smallpox and measles were still more destructive than in modern times, and recurred as frequently. St. Anthony's fire was the dread of town and country; and that disgusting disease, the leprosy, which, in consequence of the Crusades, spread its insinuating poison in all directions, snatched from the paternal hearth innumerable victims who, banished from human society, pined away in lonely huts, whither they were accompanied only by the pity of the benevolent and their own despair. All these calamities, of which the moderns have scarcely retained any recollection, were heightened to an incredible degree by the Black Death, which spread boundless devastation and misery over Italy. Men's minds were everywhere morbidly sensitive; and as it happened with individuals whose senses, when they are suffering under anxiety, become more irritable, so that trifles are magnified into objects of great alarm, and slight shocks, which would scarcely affect the spirits when in health, gave rise in them to severe diseases, so was it with this whole nation, at all times so alive to emotions, and at that period so sorely oppressed with the horrors of death.

The bite of venomous spiders, or rather the unreasonable fear of its consequences, excited at such a juncture, though it could not have done so at an earlier period, a violent nervous disorder, which, like St. Vitus's dance in Germany, spread by sympathy, increasing in severity as it took a wider range, and still further extending its ravages from its long continuance. Thus, from the middle of the fourteenth century, the furies of THE DANCE brandished their scourge over afflicted mortals; and music, for which the inhabitants of Italy, now probably for the first time, manifested susceptibility and talent, became capable of exciting ecstatic attacks in those affected, and then furnished the magical means of exorcising their melancholy.

SECT. 3 – INCREASE

At the close of the fifteenth century we find that tarantism had spread beyond the boundaries of Apulia, and that the fear of being bitten by venomous spiders had increased. Nothing short of death itself was expected from the wound which these insects inflicted, and if those who were bitten escaped with their lives, they were said to be seen pining away in a desponding state of lassitude. Many became weak-sighted or hard of hearing, some lost the power of speech, and all were insensible to ordinary causes of excitement. Nothing but the flute or the cithern afforded them relief. At the sound of these instruments they awoke as it were by enchantment, opened their eyes, and moving slowly at first, according to the measure of the music, were, as the time quickened, gradually hurried on to the most passionate dance. It was generally observable that country people, who were rude, and ignorant of music, evinced on these occasions an unusual degree of grace, as if they had been well practised in elegant movements of the body; for it is a peculiarity in nervous disorders of this kind, that the organs of motion are in an altered condition, and are completely under the control of the overstrained spirits. Cities and villages alike resounded throughout the summer season with the notes of fifes, clarinets, and Turkish drums; and patients were everywhere to be met with who looked to dancing as their only remedy. Alexander ab Alexandro, who gives this account, saw a young man in a remote village who was seized with a violent attack of tarantism. He listened with eagerness and a fixed stare to the sound of a drum, and his graceful movements gradually became more and more violent, until his dancing was converted into a succession of frantic leaps, which required the utmost exertion of his whole strength. In the midst of this overstrained exertion of mind and body the music suddenly ceased, and he immediately fell powerless to the ground, where he lay senseless and motionless until its magical effect again aroused him to a renewal of his impassioned performances.

At the period of which we are treating there was a general conviction, that by music and dancing the poison of the tarantula was distributed over the whole body, and expelled through the skin, but that if there remained the slightest vestige of it in the vessels, this became a permanent germ of the disorder, so that the dancing fits might again and again be excited ad infinitum by music. This belief, which resembled the delusion of those insane persons who, being by artful management freed from the imagined causes of their sufferings, are but for a short time released from their false notions, was attended with the most injurious effects: for in consequence of it those affected necessarily became by degrees convinced of the incurable nature of their disorder. They expected relief, indeed, but not a cure, from music; and when the heat of summer awakened a recollection of the dances of the preceding year, they, like the St. Vitus's dancers of the same period before St. Vitus's day, again grew dejected and misanthropic, until, by music and dancing, they dispelled the melancholy which had become with them a kind of sensual enjoyment.

Under such favourable circumstances, it is clear that tarantism must every year have made further progress. The number of those affected by it increased beyond all belief, for whoever had either actually been, or even fancied that he had been, once bitten by a poisonous spider or scorpion, made his appearance annually wherever the merry notes of the tarantella resounded. Inquisitive females joined the throng and caught the disease, not indeed from the poison of the spider, but from the mental poison which they eagerly received through the eye; and thus the cure of the tarantati gradually became established as a regular festival of the populace, which was anticipated with impatient delight.

Without attributing more to deception and fraud than to the peculiar nature of a progressive mental malady, it may readily be conceived that the cases of this strange disorder now grew more frequent. The celebrated Matthioli, who is worthy of entire confidence, gives his account as an eyewitness. He saw the same extraordinary effects produced by music as Alexandro, for, however tortured with pain, however hopeless of relief the patients appeared, as they lay stretched on the couch of sickness, at the very first sounds of those melodies which made an impression on them – but this was the case only with the tarantellas composed expressly for the purpose – they sprang up as if inspired with new life and spirit, and, unmindful of their disorder, began to move in measured gestures, dancing for hour together without fatigue, until, covered with a kindly perspiration, they felt a salutary degree of lassitude, which relieved them for a time at least, perhaps even for a whole year, from their defection and oppressive feeling of general indisposition. Alexandro's experience of the injurious effects resulting from a sudden cessation of the music was generally confirmed by Matthioli. If the clarinets and drums ceased for a single moment, which, as the most skilful payers were tired out by the patients, could not but happen occasionally, they suffered their limbs to fall listless, again sank exhausted to the ground, and could find no solace but in a renewal of the dance. On this account care was taken to continue the music until exhaustion was produced; for it was better to pay a few extra musicians, who might relieve each other, than to permit the patient, in the midst of this curative exercise, to relapse into so deplorable a state of suffering. The attack consequent upon the bite of the tarantula, Matthioli describes as varying much in its manner. Some became morbidly exhilarated, so that they remained for a long while without sleep, laughing, dancing, and singing in a state of the greatest excitement. Others, on the contrary, were drowsy. The generality felt nausea and suffered from vomiting, and some had constant tremors. Complete mania was no uncommon occurrence, not to mention the usual dejection of spirits and other subordinate symptoms.

SECT. 4 – IDIOSYNCRASIES – MUSIC

Unaccountable emotions, strange desires, and morbid sensual irritations of all kinds, were as prevalent as in the St. Vitus's dance and similar great nervous maladies. So late as the sixteenth century patients were seen armed with glittering swords which, during the attack, they brandished with wild gestures, as if they were going to engage in a fencing match. Even women scorned all female delicacy, and, adopting this impassioned demeanour, did the same; and this phenomenon, as well as the excitement which the tarantula dancers felt at the sight of anything with metallic lustre, was quite common up to the period when, in modern times, the disease disappeared.

The abhorrence of certain colours, and the agreeable sensations produced by others, were much more marked among the excitable Italians than was the case in the St. Vitus's dance with the more phlegmatic Germans. Red colours, which the St. Vitus's dancers detested, they generally liked, so that a patient was seldom seen who did not carry a red handkerchief for his gratification, or greedily feast his eyes on any articles of red clothing worn by the bystanders. Some preferred yellow, others black colours, of which an explanation was sought, according to the prevailing notions of the times, in the difference of temperaments. Others, again, were enraptured with green; and eyewitnesses describe this rage for colours as so extraordinary, that they can scarcely find words with which to express their astonishment. No sooner did the patients obtain a sight of the favourite colour than, new as the impression was, they rushed like infuriated animals towards the object, devoured it with their eager looks, kissed and caressed it in every possible way, and gradually resigning themselves to softer sensations, adopted the languishing expression of enamoured lovers, and embraced the handkerchief, or whatever other article it might be, which was presented to them, with the most intense ardour, while the tears streamed from their eyes as if they were completely overwhelmed by the inebriating impression on their senses.

The dancing fits of a certain Capuchin friar in Tarentum excited so much curiosity, that Cardinal Cajetano proceeded to the monastery, that he might see with his own eyes what was going on. As soon as the monk, who was in the midst of his dance, perceived the spiritual prince clothed in his red garments, he no longer listened to the tarantella of the musicians, but with strange gestures endeavoured to approach the Cardinal, as if he wished to count the very threads of his scarlet robe, and to allay his intense longing by its odour. The interference of the spectators, and his own respect, prevented his touching it, and thus the irritation of his senses not being appeased, he fell into a state of such anguish and disquietude, that he presently sank down in a swoon, from which he did not recover until the Cardinal compassionately gave him his cape. This he immediately seized in the greatest ecstasy, and pressed now to his breast, now to his forehead and cheeks, and then again commenced his dance as if in the frenzy of a love fit.

At the sight of colours which they disliked, patients flew into the most violent rage, and, like the St. Vitus's dancers when they saw red objects, could scarcely be restrained from tearing the clothes of those spectators who raised in them such disagreeable sensations.

Another no less extraordinary symptom was the ardent longing for the sea which the patients evinced. As the St. John's dancers of the fourteenth century saw, in the spirit, the heavens open and display all the splendour of the saints, so did those who were suffering under the bite of the tarantula feel themselves attracted to the boundless expanse of the blue ocean, and lost themselves in its contemplation. Some songs, which are still preserved, marked this peculiar longing, which was moreover expressed by significant music, and was excited even by the bare mention of the sea. Some, in whom this susceptibility was carried to the greatest pitch, cast themselves with blind fury into the blue waves, as the St. Vitus's dancers occasionally did into rapid rivers. This condition, so opposite to the frightful state of hydrophobia, betrayed itself in others only in the pleasure afforded them by the sight of clear water in glasses. These they bore in their hands while dancing, exhibiting at the same time strange movements, and giving way to the most extravagant expressions of their feeling. They were delighted also when, in the midst of the space allotted for this exercise, more ample vessels, filled with water, and surrounded by rushes and water plants, were placed, in which they bathed their heads and arms with evident pleasure. Others there were who rolled about on the ground, and were, by their own desire, buried up to the neck in the earth, in order to alleviate the misery of their condition; not to mention an endless variety of other symptoms which showed the perverted action of the nerves.

All these modes of relief, however, were as nothing in comparison with the irresistible charms of musical sound. Attempts had indeed been made in ancient times to mitigate the pain of sciatica, or the paroxysms of mania, by the soft melody of the flute, and, what is still more applicable to the present purpose, to remove the danger arising from the bite of vipers by the same means. This, however, was tried only to a very small extent. But after being bitten by the tarantula, there was, according to popular opinion, no way of saving life except by music; and it was hardly considered as an exception to the general rule, that every now and then the bad effects of a wound were prevented by placing a ligature on the bitten limb, or by internal medicine, or that strong persons occasionally withstood the effects of the poison, without the employment of any remedies at all. It was much more common, and is quite in accordance with the nature of so exquisite a nervous disease, to hear accounts of many who, when bitten by the tarantula, perished miserably because the tarantella, which would have afforded them deliverance, was not played to them. It was customary, therefore, so early as the commencement of the seventeenth century, for whole bands of musicians to traverse Italy during the summer months, and, what is quite unexampled either in ancient or modern times, the cure of the Tarantati in the different towns and villages was undertaken on a grand scale. This season of dancing and music was called "the women's little carnival," for it was women more especially who conducted the arrange-ments; so that throughout the whole country they saved up their spare money, for the purpose of rewarding the welcome musicians, and many of them neglected their household employments to participate in this festival of the sick. Mention is even made of one benevolent lady (Mita Lupa) who had expended her whole fortune on this object.

The music itself was of a kind perfectly adapted to the nature of the malady, and it made so deep an impression on the Italians, that even to the present time, long since the extinction of the disorder, they have retained the tarantella, as a particular species of music employed for quick, lively dancing. The different kinds of tarantella were distinguished, very significantly, by particular names, which had reference to the moods observed in the patients. Whence it appears that they aimed at representing by these tunes even the idiosyncrasies of the mind as expressed in the countenance. Thus there was one kind of tarantella which was called "Panno rosso," a very lively, impassioned style of music, to which wild dithyrambic songs were adapted; another, called "Panno verde," which was suited to the milder excitement of the senses caused by green colours, and set to Idyllian songs of verdant fields and shady groves. A third was named "Cinque tempi:" a fourth "Moresca," which was played to a Moorish dance; a fifth, "Catena;" and a sixth, with a very appropriate designation, "Spallata," as if it were only fit to be played to dancers who were lame in the shoulder. This was the slowest and least in vogue of all. For those who loved water they took care to select love songs, which were sung to corresponding music, and such persons delighted in hearing of gushing springs and rushing cascades and streams. It is to be regretted that on this subject we are unable to give any further information, for only small fragments of songs, and a very few tarantellas, have been preserved which belong to a period so remote as the beginning of the seventeenth, or at furthest the end of the sixteenth century.

The music was almost wholly in the Turkish style (aria Turchesca), and the ancient songs of the peasantry of Apulia, which increased in number annually, were well suited to the abrupt and lively notes of the Turkish drum and the shepherd's pipe. These two instruments were the favourites in the country, but others of all kinds were played in towns and villages, as an accompaniment to the dances of the patients and the songs of the spectators. If any particular melody was disliked by those affected, they indicated their displeasure by violent gestures expressive of aversion. They could not endure false notes, and it is remarkable that uneducated boors, who had never in their lives manifested any perception of the enchanting power of harmony, acquired, in this respect, an extremely refined sense of hearing, as if they had been initiated into the profoundest secrets of the musical art. It was a matter of every day's experience, that patients showed a predilection for certain tarantellas, in preference to others, which gave rise to the composition of a great variety of these dances. They were likewise very capricious in their partialities for particular instruments; so that some longed for the shrill notes of the trumpet, others for the softest music produced by the vibration of strings.

Tarantism was at its greatest height in Italy in the seventeenth century, long after the St. Vitus's Dance of Germany had disappeared. It was not the natives of the country only who were attacked by this complaint. Foreigners of every colour and of every race, negroes, gipsies, Spaniards, Albanians, were in like manner affected by it. Against the effects produced by the tarantula's bite, or by the sight of the sufferers, neither youth nor age afforded any protection; so that even old men of ninety threw aside their crutches at the sound of the tarantella, and, as if some magic potion, restorative of youth and vigour, were flowing through their veins, joined the most extravagant dancers. Ferdinando saw a boy five years old seized with the dancing mania, in consequence of the bite of a tarantula, and, what is almost past belief, were it not supported by the testimony of so credible an eyewitness, even deaf people were not exempt from this disorder, so potent in its effect was the very sight of those affected, even without the exhilarating emotions caused by music.

Subordinate nervous attacks were much more frequent during this century than at any former period, and an extraordinary icy coldness was observed in those who were the subject of them; so that they did not recover their natural heat until they had engaged in violent dancing. Their anguish and sense of oppression forced from them a cold perspiration; the secretion from the kidneys was pale, and they had so great a dislike to everything cold, that when water was offered them they pushed it away with abhorrence. Wine, on the contrary, they all drank willingly, without being heated by it, or in the slightest degree intoxicated. During the whole period of the attack they suffered from spasms in the stomach, and felt a disinclination to take food of any kind. They used to abstain some time before the expected seizures from meat and from snails, which they thought rendered them more severe, and their great thirst for wine may therefore in some measure be attributable to the want of a more nutritious diet; yet the disorder of the nerves was evidently its chief cause, and the loss of appetite, as well as the necessity for support by wine, were its effects. Loss of voice, occasional blindness, vertigo, complete insanity, with sleeplessness, frequent weeping without any ostensible cause, were all usual symptoms. Many patients found relief from being placed in swings or rocked in cradles; others required to be roused from their state of suffering by severe blows on the soles of their feet; others beat themselves, without any intention of making a display, but solely for the purpose of allaying the intense nervous irritation which they felt; and a considerable number were seen with their bellies swollen, like those of the St. John's dancers, while the violence of the intestinal disorder was indicated in others by obstinate constipation or diarrhoea and vomiting. These pitiable objects gradually lost their strength and their colour, and creeping about with injected eyes, jaundiced complexions, and inflated bowels, soon fell into a state of profound melancholy, which found food and solace in the solemn tolling of the funeral bell, and in an abode among the tombs of cemeteries, as is related of the Lycanthropes of former times.

The persuasion of the inevitable consequences of being bitten by the tarantula, exercised a dominion over men's minds which even the healthiest and strongest could not shake off. So late as the middle of the sixteenth century, the celebrated Fracastoro found the robust bailiff of his landed estate groaning, and, with the aspect of a person in the extremity of despair, suffering the very agonies of death from a sting in the neck, inflicted by an insect which was believed to be a tarantula. He kindly administered without delay a potion of vinegar and Armenian bole, the great remedy of those days for the plague of all kinds of animal poisons, and the dying man was, as if by a miracle, restored to life and the power of speech. Now, since it is quite out of the question that the bole could have anything to do with the result in this case, notwithstanding Fracastoro's belief in its virtues, we can only account for the cure by supposing, that a confidence in so great a physician prevailed over this fatal disease of the imagination, which would otherwise have yielded to scarcely any other remedy except the tarantella. Ferdinando was acquainted with women who, for thirty years in succession, had overcome the attacks of this disorder by a renewal of their annual dance – so long did they maintain their belief in the yet undestroyed poison of the tarantula's bite, and so long did that mental affection continue to exist, after it had ceased to depend on any corporeal excitement.

Wherever we turn, we find that this morbid state of mind prevailed, and was so supported by the opinions of the age, that it needed only a stimulus in the bite of the tarantula, and the supposed certainty of its very disastrous consequences, to originate this violent nervous disorder. Even in Ferdinando's time there were many who altogether denied the poisonous effects of the tarantula's bite, whilst they considered the disorder, which annually set Italy in commotion, to be a melancholy depending on the imagination. They dearly expiated this scepticism, however, when they were led, with an inconsiderate hardihood, to test their opinions by experiment; for many of them became the subjects of severe tarantism, and even a distinguished prelate, Jo. Baptist Quinzato, Bishop of Foligno, having allowed himself, by way of a joke, to be bitten by a tarantula, could obtain a cure in no other way than by being, through the influence of the tarantella, compelled to dance. Others among the clergy, who wished to shut their ears against music, because they considered dancing derogatory to their station, fell into a dangerous state of illness by thus delaying the crisis of the malady, and were obliged at last to save themselves from a miserable death by submitting to the unwelcome but sole means of cure. Thus it appears that the age was so little favourable to freedom of thought, that even the most decided sceptics, incapable of guarding themselves against the recollection of what had been presented to the eye, were subdued by a poison, the powers of which they had ridiculed, and which was in itself inert in its effect.

SECT. 5 – HYSTERIA

Different characteristics of the morbidly excited vitality having been rendered prominent by tarantism in different individuals, it could not but happen that other derangements of the nerves would assume the form of this whenever circumstances favoured such a transition. This was more especially the case with hysteria, that proteiform and mutable disorder, in which the imaginations, the superstitions, and the follies of all ages have been evidently reflected. The "Carnevaletto delle Donne" appeared most opportunely for those who were hysterical. Their disease received from it, as it had at other times from other extraordinary customs, a peculiar direction; so that, whether bitten by the tarantula or not, they felt compelled to participate in the dances of those affected, and to make their appearance at this popular festival, where they had an opportunity of triumphantly exhibiting their sufferings. Let us here pause to consider the kind of life which the women in Italy led. Lonely, and deprived by cruel custom of social intercourse, that fairest of all enjoyments, they dragged on a miserable existence. Cheerfulness and an inclination to sensual pleasures passed into compulsory idleness, and, in many, into black despondency. Their imaginations became disordered – a pallid countenance and oppressed respiration bore testimony to their profound sufferings. How could they do otherwise, sunk as they were in such extreme misery, than seize the occasion to burst forth from their prisons and alleviate their miseries by taking part in the delights of music? Nor should we here pass unnoticed a circumstance which illustrates, in a remarkable degree, the psychological nature of hysterical sufferings, namely, that many chlorotic females, by joining the dancers at the Carnevaletto, were freed from their spasms and oppression of breathing for the whole year, although the corporeal cause of their malady was not removed. After such a result, no one could call their self-deception a mere imposture, and unconditionally condemn it as such.

This numerous class of patients certainly contributed not a little to the maintenance of the evil, for their fantastic sufferings, in which dissimulation and reality could scarcely be distinguished even by themselves, much less by their physicians, were imitated in the same way as the distortions of the St. Vitus's dancers by the impostors of that period. It was certainly by these persons also that the number of subordinate symptoms was increased to an endless extent, as may be conceived from the daily observation of hysterical patients who, from a morbid desire to render themselves remarkable, deviate from the laws of moral propriety. Powerful sexual excitement had often the most decided influence over their condition. Many of them exposed themselves in the most indecent manner, tore their hair out by the roots, with howling and gnashing of their teeth; and when, as was sometimes the case, their unsatisfied passion hurried them on to a state of frenzy, they closed their existence by self destruction; it being common at that time for these unfortunate beings to precipitate themselves into the wells.

It might hence seem that, owing to the conduct of patients of this description, so much of fraud and falsehood would be mixed up with the original disorder that, having passed into another complaint, it must have been itself destroyed. This, however, did not happen in the first half of the seventeenth century; for, as a clear proof that tarantism remained substantially the same and quite unaffected by hysteria, there were in many places, and in particular at Messapia, fewer women affected than men, who, in their turn, were in no small proportion led into temptation by sexual excitement. In other places, as, for example, at Brindisi, the case was reversed, which may, as in other complaints, be in some measure attributable to local causes. Upon the whole it appears, from concurrent accounts, that women by no means enjoyed the distinction of being attacked by tarantism more frequently than men.

It is said that the cicatrix of the tarantula bite, on the yearly or half-yearly return of the fit, became discoloured, but on this point the distinct testimony of good observers is wanting to deprive the assertion of its utter improbability.

It is not out of place to remark here that, about the same time that tarantism attained its greatest height in Italy, the bite of venomous spiders was more feared in distant parts of Asia likewise than it had ever been within the memory of man. There was this difference, however – that the symptoms supervening on the occurrence of this accident were not accompanied by the Apulian nervous disorder, which, as has been shown in the foregoing pages, had its origin rather in the melancholic temperament of the inhabitants of the south of Italy than in the nature of the tarantula poison itself. This poison is therefore, doubtless, to be considered only as a remote cause of the complaint, which, but for that temperament, would be inadequate to its production. The Persians employed a very rough means of counteracting the bad consequences of a poison of this sort. They drenched the wounded person with milk, and then, by a violent rotatory motion in a suspended box, compelled him to vomit.

SECT. 6 – DECREASE

The Dancing Mania, arising from the tarantula bite, continued with all those additions of self-deception and of the dissimulation which is such a constant attendant on nervous disorders of this kind, through the whole course of the seventeenth century. It was indeed, gradually on the decline, but up to the termination of this period showed such extraordinary symptoms that Baglivi, one of the best physicians of that time, thought he did a service to science by making them the subject of a dissertation. He repeats all the observations of Ferdinando, and supports his own assertions by the experience of his father, a physician at Lecce, whose testimony, as an eyewitness, may be admitted as unexceptionable.

The immediate consequences of the tarantula bite, the supervening nervous disorder, and the aberrations and fits of those who suffered from hysteria, he describes in a masterly style, not does he ever suffer his credulity to diminish the authenticity of his account, of which he has been unjustly accused by later writers.

Finally, tarantism has declined more and more in modern times, and is now limited to single cases. How could it possibly have maintained itself unchanged in the eighteenth century, when all the links which connected it with the Middle Ages had long since been snapped asunder? Imposture grew more frequent, and wherever the disease still appeared in its genuine form, its chief cause, namely, a peculiar cast of melancholy, which formerly had been the temperament of thousands, was now possessed only occasionally by unfortunate individuals. It might, therefore, not unreasonably be maintained that the tarantism of modern times bears nearly the same relation to the original malady as the St. Vitus's dance which still exists, and certainly has all along existed, bears, in certain cases, to the original dancing mania of the dancers of St. John.

To conclude. Tarantism, as a real disease, has been denied in toto, and stigmatised as an imposition by most physicians and naturalists, who in this controversy have shown the narrowness of their views and their utter ignorance of history. In order to support their opinion they have instituted some experiments apparently favourable to it, but under circumstances altogether inapplicable, since, for the most part, they selected as the subjects of them none but healthy men, who were totally uninfluenced by a belief in this once so dreaded disease. From individual instances of fraud and dissimulation, such as are found in connection with most nervous affections without rendering their reality a matter of any doubt, they drew a too hasty conclusion respecting the general phenomenon, of which they appeared not to know that it had continued for nearly four hundred years, having originated in the remotest periods of the Middle Ages. The most learned and the most acute among these sceptics is Serao the Neapolitan. His reasonings amount to this, that he considers the disease to be a very marked form of melancholia, and compares the effect of the tarantula bite upon it to stimulating with spurs a horse which is already running. The reality of that effect he thus admits, and, therefore, directly confirms what in appearance only he denies. By shaking the already vacillating belief in this disorder he is said to have actually succeeded in rendering it less frequent, and in setting bounds to imposture; but this no more disproves the reality of its existence than the oft repeated detection of imposition has been able in modern times to banish magnetic sleep from the circle of natural phenomena, though such detection has, on its side, rendered more rare the incontestable effects of animal magnetism. Other physicians and naturalists have delivered their sentiments on tarantism, but as they have not possessed an enlarged knowledge of its history their views do not merit particular exposition. It is sufficient for the comprehension of everyone that we have presented the facts from all extraneous speculation.

CHAPTER III

THE DANCING MANIA IN ABYSSINIA

***

SECT. 1 – TIGRETIER

Both the St. Vitus's dance and tarantism belonged to the ages in which they appeared. They could not have existed under the same latitude at any other epoch, for at no other period were the circumstances which prepared the way for them combined in a similar relation to each other, and the mental as well as corporeal temperaments of nations, which depend on causes such as have been stated, are as little capable of renewal as the different stages of life in individuals. This gives so much the more importance to a disease but cursorily alluded to in the foregoing pages, which exists in Abyssinia, and which nearly resembles the original mania of the St. John's dancers, inasmuch as it exhibits a perfectly similar ecstasy, with the same violent effect on the nerves of motion. It occurs most frequently in the Tigre country, being thence call Tigretier, and is probably the same malady which is called in Ethiopian language Astaragaza. On this subject we will introduce the testimony of Nathaniel Pearce, an eyewitness, who resided nine years in Abyssinia. "The Tigretier," he says he, "is more common among the women than among the men. It seizes the body as if with a violent fever, and from that turns to a lingering sickness, which reduces the patients to skeletons, and often kills them if the relations cannot procure the proper remedy. During this sickness their speech is changed to a kind of stuttering, which no one can understand but those afflicted with the same disorder. When the relations find the malady to be the real tigretier, they join together to defray the expense of curing it; the first remedy they in general attempt is to procure the assistance of a learned Dofter, who reads the Gospel of St. John, and drenches the patient with cold water daily for the space of seven days, an application that very often proves fatal. The most effectual cure, though far more expensive than the former, is as follows:- The relations hire for a certain sum of money a band of trumpeters, drummers, and fifers, and buy a quantity of liquor; then all the young men and women of the place assemble at the patient's house to perform the following most extraordinary ceremony.

"I was once called in by a neighbour to see his wife, a very young woman, who had the misfortune to be afflicted with this disorder; and the man being an old acquaintance of mine, and always a close comrade in the camp, I went every day, when at home, to see her, but I could not be of any service to her, though she never refused my medicines. At this time I could not understand a word she said, although she talked very freely, nor could any of her relations understand her. She could not bear the sight of a book or a priest, for at the sight of either she struggled, and was apparently seized with acute agony, and a flood of tears, like blood mingled with water, would pour down her face from her eyes. She had lain three months in this lingering state, living upon so little that it seemed not enough to keep a human body alive; at last her husband agreed to employ the usual remedy, and, after preparing for the maintenance of the band during the time it would take to effect the cure, he borrowed from all his neighbours their silver ornaments, and loaded her legs, arms and neck with them.

"The evening that the band began to play I seated myself close by her side as she lay upon the couch, and about two minutes after the trumpets had begun to sound I observed her shoulders begin to move, and soon afterwards her head and breast, and in less than a quarter of an hour she sat upon her couch. The wild look she had, though sometimes she smiled, made me draw off to a greater distance, being almost alarmed to see one nearly a skeleton move with such strength; her head, neck, shoulders, hands and feet all made a strong motion to the sound of the music, and in this manner she went on by degrees, until she stood up on her legs upon the floor. Afterwards she began to dance, and at times to jump about, and at last, as the music and noise of the singers increased, she often sprang three feet from the ground. When the music slackened she would appear quite out of temper, but when it became louder she would smile and be delighted. During this exercise she never showed the least symptom of being tired, though the musicians were thoroughly exhausted; and when they stopped to refresh themselves by drinking and resting a little she would discover signs of discontent.

"Next day, according to the custom in the cure of this disorder, she was taken into the marketplace, where several jars of maize or tsug were set in order by the relations, to give drink to the musicians and dancers. When the crowd had assembled, and the music was ready, she was brought forth and began to dance and throw herself into the maddest postures imaginable, and in this manner she kept on the whole day.

Towards evening she began to let fall her silver ornaments from her neck, arms, and legs, one at a time, so that in the course of three hours she was stripped of every article. A relation continually kept going after her as she danced, to pick up the ornaments, and afterwards delivered them to the owners from whom they were borrowed. As the sun went down she made a start with such swiftness that the fastest runner could not come up with her, and when at the distance of about two hundred yards she dropped on a sudden as if shot. Soon afterwards a young man, on coming up with her, fired a matchlock over her body, and struck her upon the back with the broad side of his large knife, and asked her name, to which she answered as when in her common senses – a sure proof of her being cured; for during the time of this malady those afflicted with it never answer to their Christian names. She was now taken up in a very weak condition and carried home, and a priest came and baptised her again in the name of the Father, Son, and Holy

Ghost, which ceremony concluded her cure. Some are taken in this manner to the marketplace for many days before they can be cured, and it sometimes happens that they cannot be cured at all. I have seen them in these fits dance with a BRULY, or bottle of maize, upon their heads without spilling the liquor, or letting the bottle fall, although they have put themselves into the most extravagant postures.

"I could not have ventured to write this from hearsay, nor could I conceive it possible, until I was obliged to put this remedy in practice upon my own wife, who was seized with the same disorder, and then I was compelled to have a still nearer view of this strange disorder. I at first thought that a whip would be of some service, and one day attempted a few strokes when unnoticed by any person, we being by ourselves, and I having a strong suspicion that this ailment sprang from the weak minds of women, who were encouraged in it for the sake of the grandeur, rich dress, and music which accompany the cure.

But how much was I surprised, the moment I struck a light blow, thinking to do good, to find that she became like a corpse, and even the joints of her fingers became so stiff that I could not straighten them; indeed, I really thought that she was dead, and immediately made it known to the people in the house that she had fainted, but did not tell them the cause, upon which they immediately brought music, which I had for many days denied them, and which soon revived her; and I then left the house to her relations to cure her at my expense, in the manner I have before mentioned, though it took a much longer time to cure my wife than the woman I have just given an account of. One day I went privately, with a companion, to see my wife dance, and kept at a short distance, as I was ashamed to go near the crowd. On looking steadfastly upon her, while dancing or jumping, more like a deer than a human being, I said that it certainly was not my wife; at which my companion burst into a fit of laughter, from which he could scarcely refrain all the way home. Men are sometimes afflicted with this dreadful disorder, but not frequently. Among the Amhara and Galla it is not so common."

Such is the account of Pearce, who is every way worthy of credit, and whose lively description renders the traditions of former times respecting the St. Vitus's dance and tarantism intelligible, even to those who are sceptical respecting the existence of a morbid state of the mind and body of the kind described, because, in the present advanced state of civilisation among the nations of Europe, opportunities for its development no longer occur. The credibility of this energetic but by no means ambitious man is not liable to the slightest suspicion, for, owing to his want of education, he had no knowledge of the phenomena in question, and his work evinces throughout his attractive and unpretending impartiality.

Comparison is the mother of observation, and may here elucidate one phenomenon by another – the past by that which still exists. Oppression, insecurity, and the influence of a very rude priestcraft, are the powerful causes which operated on the Germans and Italians of the Middle Ages, as they now continue to operate on the Abyssinians of the present day. However these people may differ from us in their descent, their manners and their customs, the effects of the above mentioned causes are the same in Africa as they were in Europe, for they operate on man himself independently of the particular locality in which he may be planted; and the conditions of the Abyssinians of modern times is, in regard to superstition, a mirror of the condition of the European nations of the middle ages. Should this appear a bold assertion it will be strengthened by the fact that in Abyssinia two examples of superstitions occur which are completely in accordance with occurrences of the Middle Ages that took place contemporarily with the dancing mania. THE ABYSSINIANS HAVE THEIR CHRISTIAN FLAGELLANTS, AND THERE EXISTS AMONG THEM A BELIEF IN A ZOOMORPHISM, WHICH

PRESENTS A LIVELY IMAGE OF THE LYCANTHROPY OF THE MIDDLE AGES. Their flagellants are called Zackarys. They are united into a separate Christian fraternity, and make their processions through the towns and villages with great noise and tumult, scourging themselves till they draw blood, and wounding themselves with knives. They boast that they are descendants of St. George. It is precisely in Tigre, the country of the Abyssinian dancing mania, where they are found in the greatest numbers, and where they have, in the neighbourhood of Axum, a church of their own, dedicated to their patron saint, Oun Arvel. Here there is an ever-burning lamp, and they contrive to impress a belief that this is kept alight by supernatural means. They also here keep a holy water, which is said to be a cure for those who are affected by the dancing mania.

The Abyssinian Zoomorphism is a no less important phenomenon, and shows itself a manner quite peculiar. The blacksmiths and potters form among the

Abyssinians a society or caste called in Tigre TEBBIB, and in Amhara BUDA, which is held in some degree of contempt, and excluded from the sacrament of the Lord's Supper, because it is believed that they can change themselves into hyaenas and other beasts of prey, on which account they are feared by everybody, and regarded with horror. They artfully contrive to keep up this superstition, because by this separation they preserve a monopoly of their lucrative trades, and as in other respects they are good Christians (but few Jews or Mahomedans live among them), they seem to attach no great consequence to their excommunication. As a badge of distinction they wear a golden earring, which is frequently found in the ears of Hyaenas that are killed, without its having ever been discovered how they catch these animals, so as to decorate them with this strange ornament, and this removes in the minds of the people all doubt as to the supernatural powers of the smiths and potters. To the Budas is also ascribed the gift of enchantment, especially that of the influence of the evil eye. They nevertheless live unmolested, and are not condemned to the flames by fanatical priests, as the lycanthropes were in the Middle Ages.

CHAPTER IV

SYMPATHY

***

Imitation – compassion – sympathy, these are imperfect designations for a common bond of union among human beings – for an instinct which connects individuals with the general body, which embraces with equal force reason and folly, good and evil, and diminishes the praise of virtue as well as the criminality of vice. In this impulse there are degrees, but no essential differences, from the first intellectual efforts of the infant mind, which are in a great measure based on imitation, to that morbid condition of the soul in which the sensible impression of a nervous malady fetters the mind, and finds its way through the eye directly to the diseased texture, as the electric shock is propagated by contact from body to body. To this instinct of imitation, when it exists in its highest degree, is united a loss of all power over the will, which occurs as soon as the impression on the senses has become firmly established, producing a condition like that of small animals when they are fascinated by the look of a serpent. By this mental bondage morbid sympathy is clearly and definitely distinguished from all subordinate degrees of this instinct, however closely allied the imitation of a disorder may seem to be to that of a mere folly, of an absurd fashion, of an awkward habit in speech and manner, or even of a confusion of ideas. Even these latter imitations, however, directed as they are to foolish and pernicious objects, place the self-independence of the greater portion of mankind in a very doubtful light, and account for their union into a social whole. Still more nearly allied to morbid sympathy than the imitation of enticing folly, although often with a considerable admixture of the latter, is the diffusion of violent excitements, especially those of a religious or political character, which have so powerfully agitated the nations of ancient and modern times, and which may, after an incipient compliance, pass into a total loss of power over the will, and an actual disease of the mind. Far be it from us to attempt to awaken all the various tones of this chord, whose vibrations reveal the profound secrets which lie hid in the inmost recesses of the soul. We might well want powers adequate to so vast an undertaking. Our business here is only with that morbid sympathy by the aid of which the dancing mania of the Middle Ages grew into a real epidemic. In order to make this apparent by comparison, it may not be out of place, at the close of this inquiry, to introduce a few striking examples:–

1) "At a cotton manufactory at Hodden Bridge, in Lancashire, a girl, on the fifteenth of February, 1787, put a mouse into the bosom of another girl, who had a great dread of mice. The girl was immediately thrown into a fit, and continued in it, with the most violent convulsions, for twenty-four hours. On the following day three more girls were seized in the same manner, and on the 17th six more. By this time the alarm was so great

that the whole work, in which 200 or 300 were employed, was totally stopped, and an idea prevailed that a particular disease had been introduced by a bag of cotton opened in the house. On Sunday the 18th, Dr. St. Clare was sent for from Preston; before he arrived three more were seized, and during that night and the morning of the 19th, eleven more, making in all twenty-four. Of these, twenty-one were young women, two were girls of about ten years of age, and one man, who had been much fatigued with holding the girls. Three of the number lived about two miles from the place where the disorder first broke out, and three at another factory at Clitheroe, about five miles distant, which last and two more were infected entirely from report, not having seen the other patients, but, like them and the rest of the country, strongly impressed with the idea of the plague being caught from the cotton. The symptoms were anxiety, strangulation, and very strong convulsions; and these were so violent as to last without any intermission from a quarter of an hour to twenty-four hours, and to require four or five persons to prevent the patients from tearing their hair and dashing their heads against the floor or walls. Dr. St. Clare had taken with him a portable electrical machine, and by electric shocks the patients were universally relieved without exception. As soon as the patients and the country were assured that the complaint was merely nervous, easily cured, and not introduced by the cotton, no fresh person was affected. To dissipate their apprehensions still further, the best effects were obtained by causing them to take a cheerful glass and join in a dance. On Tuesday the 20th, they danced, and the next day were all at work, except two or three, who were much weakened by their fits."

The occurrence here described is remarkable on this account, that there was no important predisposing cause for convulsions in these young women, unless we consider as such their miserable and confined life in the workrooms of a spinning manufactory. It did not arise from enthusiasm, nor is it stated that the patients had been the subject of any other nervous disorders. In another perfectly analogous case, those attacked were all suffering from nervous complaints, which roused a morbid sympathy in them at the sight of a person seized with convulsions. This, together with the supervention of hysterical fits, may aptly enough be compared to tarantism.

2) "A young woman of the lowest order, twenty-one years of age, and of a strong frame, came on the 13th of January, 1801, to visit a patient in the Charite Hospital at Berlin, where she had herself been previously under treatment for an inflammation of the chest with tetanic spasms, and immediately on entering the ward, fell down in strong convulsions. At the sight of her violent contortions six other female patients immediately became affected in the

same way, and by degrees eight more were in like manner attacked with strong convulsions. All these patients were from sixteen to twenty-five years of age, and suffered without exception, one from spasms in the stomach, another from palsy, a third from lethargy, a fourth from fits with consciousness, a fifth from catalepsy, a sixth from syncope, &c. The convulsions, which alternated in various ways with tonic spasms, were accompanied by loss of sensibility, and were invariably preceded by languor with heavy sleep, which was followed by the fits in the course of a minute or two; and it is remarkable that in all these patients their former nervous disorders, not excepting paralysis, disappeared, returning, however, after the subsequent removal of their new complaint. The treatment, during the course of which two of the nurses, who were young women, suffered similar attacks, was continued for four months. It was finally successful, and consisted principally in the administration of opium, at that time the favourite remedy.

Now every species of enthusiasm, every strong affection, every violent passion, may lead to convulsions – to mental disorders – to a concussion of the nerves, from the sensorium to the very finest extremities of the spinal chord. The whole world is full of examples of this afflicting state of turmoil, which, when the mind is carried away by the force of a sensual impression that destroys its freedom, is irresistibly propagated by imitation. Those who are thus infected do not spare even their own lives, but as a hunted flock of sheep will follow their leader and rush over a precipice, so will whole hosts of enthusiasts, deluded by their infatuation, hurry on to a self-inflicted death. Such has ever been the case, from the days of the Milesian virgins to the modern associations for self-destruction. Of all enthusiastic infatuations, however, that of religion is the most fertile in disorders of the mind as well as of the body, and both spread with the greatest facility by sympathy. The history of the Church furnishes innumerable proofs of this, but we need go no further than the most recent times.

3) In a methodist chapel at Redruth, a man during divine service cried out with a loud voice, "What shall I do to be saved?" at the same time manifesting the greatest uneasiness and solicitude respecting the condition of his soul. Some other members of the congregation, following his example, cried out in the same form of words, and seemed shortly after to suffer the most excruciating bodily pain. This strange occurrence was soon publicly known, and hundreds of people who had come thither, either attracted by curiosity or a desire from other motives to see the sufferers, fell into the same state. The chapel remained open for some days and nights, and from that point the new disorder spread itself, with the rapidity of lightning, over the neighbouring towns of Camborne, Helston, Truro, Penryn and Falmouth, as well

as over the villages in the vicinity. Whilst thus advancing, it decreased in some measure at the place where it had first appeared, and it confined itself throughout to the Methodist chapels. It was only by the words which have been mentioned that it was excited, and it seized none but people of the lowest education. Those who were attacked betrayed the greatest anguish, and fell into convulsions; others cried out, like persons possessed, that the Almighty would straightway pour out His wrath upon them, that the wailings of tormented spirits rang in their ears, and that they saw hell open to receive them. The clergy, when in the course of their sermons they perceived that persons were thus seized, earnestly exhorted them to confess their sins, and zealously endeavoured to convince them that they were by nature enemies to Christ; that the anger of God had therefore fallen upon them; and that if death should surprise them in the midst of their sins the eternal torments of hell would be their portion. The overexcited congregation upon this repeated their words, which naturally must have increased the fury of their convulsive attacks. When the discourse had produced its full effect the preacher changed his subject; reminded those who were suffering of the power of the Saviour, as well as of the grace of God, and represented to them in glowing colours the joys of heaven. Upon this a remarkable reaction sooner or later took place. Those who were in convulsions felt themselves raised from the lowest depths of misery and despair to the most exalted bliss, and triumphantly shouted out that their bonds were loosed, their sins were forgiven, and that they were translated to the wonderful freedom of the children of God. In the meantime their convulsions continued, and they remained during this condition so abstracted from every earthly thought that they stayed two and sometimes three days and nights together in the chapels, agitated all the time by spasmodic movements,

> and taking neither repose nor nourishment. According to a moderate computation, 4,000 people were, within a very short time, affected with this convulsive malady.

The course and symptoms of the attacks were in general as follows:– There came on at first a feeling of faintness, with rigour and a sense of weight at the pit of the stomach, soon after which the patient cried out, as if in the agonies of death or the pains of labour. The convulsions then began, first showing themselves in the muscles of the eyelids, though the eyes themselves were fixed and staring. The most frightful contortions of the countenance followed, and the convulsions now took their course downwards, so that the muscles of the neck and trunk were affected, causing a sobbing respiration, which was performed with great effort. Tremors and agitation ensued, and the patients screamed out violently, and tossed their heads about from side to side. As the complaint increased it seized the arms, and its victims beat their breasts, clasped their hands, and made all sorts of strange gestures. The observer who gives this account remarked that the lower extremities were in no instance affected. In some cases exhaustion came on in a very few minutes, but the attack usually lasted much longer, and there were even cases in which it was known to continue for sixty or seventy hours. Many of those who happened to be seated when the attack commenced bent their bodies rapidly backwards and forwards during its continuance, making a corresponding motion with their arms, like persons sawing wood. Others shouted aloud, leaped about, and threw their bodies into every possible posture, until they had exhausted their strength. Yawning took place at the commencement in all cases, but as the violence of the disorder increased the circulation and respiration became accelerated, so that the countenance assumed a swollen and puffed appearance. When exhaustion came on patients usually fainted, and remained in a stiff and motionless state until their recovery. The disorder completely resembled the St. Vitus's dance, but the fits sometimes went on to an extraordinarily violent extent, so that the author of the account once saw a woman who was seized with these convulsions resist the endeavours of four or five strong men to restrain her. Those patients who did not lose their consciousness were in general made more furious by every attempt to quiet them by force, on which account they were in general suffered to continue unmolested until nature herself brought on exhaustion. Those affected complained more or less of debility after the attacks, and cases sometimes occurred in which they passed into other disorders; thus some fell into a state of melancholy, which, however, in consequence of their religious ecstasy, was distinguished by the absence of fear and despair; and in one patient inflammation of the brain is said to have taken place. No sex or age was exempt from this epidemic malady. Children five years old and octogenarians were alike affected by it, and even men of the most powerful frame were subject to its influence. Girls and young women, however, were its most frequent victims.

4) For the last hundred years a nervous affection of a perfectly similar kind has existed in the Shetland Islands, which furnishes a striking example, perhaps the only one now existing, of the very lasting propagation by sympathy of this species of disorders. The origin of the malady was very insignificant. An epileptic woman had a fit in church, and whether it was that the minds of the congregation were excited by devotion, or that, being overcome at the sight of the strong convulsions, their sympathy was called forth, certain it is that many adult women, and even children, some of whom were of the male sex, and not more than six years old, began to complain forthwith of palpitation, followed by faintness, which passed into a motionless and apparently cataleptic condition. These symptoms lasted more than an hour, and probably recurred frequently. In the course of time, how-

ever, this malady is said to have undergone a modification, such as it exhibits at the present day. Women whom it has attacked will suddenly fall down, toss their arms about, writhe their bodies into various shapes, move their heads suddenly from side to side, and with eyes fixed and staring, utter the most dismal cries. If the fit happen on any occasion of pubic diversion, they will, as soon as it has ceased, mix with their companions and continue their amusement as if nothing had happened. Paroxysms of this kind used to prevail most during the warm months of summer, and about fifty years ago there was scarcely a Sabbath in which they did not occur. Strong passions of the mind, induced by religious enthusiasm, are also exciting causes of these fits, but like all such false tokens of divine workings, they are easily encountered by producing in the patient a different frame of mind, and especially by exciting a sense of shame: thus those affected are under the control of any sensible preacher, who knows how to "administer to a mind diseased," and to expose the folly of voluntarily yielding to a sympathy so easily resisted, or of inviting such attacks by affectation. An intelligent and pious minister of Shetland informed the physician, who gives an account of this disorder as an eyewitness, that being considerably annoyed on his first introduction into the country by these paroxysms, whereby the devotions of the church were much impeded, he obviated their repetition by assuring his parishioners that no treatment was more effectual than immersion in cold water; and as his kirk was fortunately contiguous to a freshwater lake, he gave notice that attendants should be at hand during divine service to ensure the proper means of cure. The sequel need scarcely be told. The fear of being carried out of the church, and into the water, acted like a charm; not a single Naiad was made, and the worthy minister for many years had reason to boast of one of the best regulated congregations in Scotland. As the physician above alluded to was attending divine service in the kirk of Baliasta, on the Isle of Unst, a female shriek, the indication of a convulsion fit, was heard; the minister, Mr. Ingram, of Fetlar, very properly stopped his discourse until the disturber was removed; and after advising all those who thought they might be similarly affected to leave the church, he gave out in the meantime a psalm. The congregation was thus preserved from further interruption; yet the effect of sympathy was not prevented, for as the narrator of the account was leaving the church he saw several females writhing and tossing about their arms on the green grass, who durst not, for fear of a censure from the pulpit, exhibit themselves after this manner within the sacred walls of the kirk.

In the production of this disorder, which no doubt still exists, fanaticism certainly had a smaller share than the irritable state of women out of health, who only needed excitement, no matter of what kind, to throw them into prevailing nervous paroxysms. When, however, that powerful cause of nervous disorders takes the lead, we find far more remarkable symptoms developed, and it then depends on the mental condition of the people among whom they appear whether in their spread they shall take a narrow or an extended range – whether confined to some small knot of zealots they are to vanish without a trace, or whether they are to attain even historical importance.

5) The appearance of the Convulsionnaires in France, whose inhabitants, from the greater mobility of their blood, have in general been the less liable to fanaticism, is in this respect instructive and worthy of attention. In the year 1727 there died in the capital of that country the Deacon Paris, a zealous opposer of the Ultramontanists, division having arisen in the French Church on account of the bull "Unigenitus." People made frequent visits to his tomb in the

cemetery of St. Medard, and four years afterwards (in September, 1731) a rumour was spread that miracles took place there. Patients were seized with convulsions and tetanic spasms, rolled upon the ground like persons possessed, were thrown into violent contortions of their heads and limbs, and suffered the greatest oppression, accompanied by quickness and irregularity of pulse. This novel occurrence excited the greatest sensation all over Paris, and an immense concourse of people resorted daily to the above-named cemetery in order to see so wonderful a spectacle, which the Ultramontanists immediately interpreted as a work of Satan, while their opponents ascribed it to a divine influence. The disorder soon increased, until it produced, in nervous women, clairvoyance (Schlafwachen), a phenomenon till then unknown; for one female especially attracted attention, who, blindfold, and, as it was believed, by means of the sense of smell, read every writing that was placed before her, and distinguished the characters of unknown persons. The very earth taken from the grave of the Deacon was soon thought to possess miraculous power. It was sent to numerous sick persons at a distance, whereby they were said to have been cured, and thus this nervous disorder spread far beyond the limits of the capital, so that at one time it was computed that there were more than eight hundred decided Convulsionnaires, who would hardly have increased so much in numbers had not Louis XV directed that the cemetery should be closed. The disorder itself assumed various forms, and augmented by its attacks the general excitement. Many persons, besides suffering from the convulsions, became the subjects of violent pain, which required the assistance of their brethren of the faith. On this account they, as well as those who afforded them aid, were called by the common title of Secourists. The modes of relief adopted were remarkably in accordance with those which were administered to the

St. John's dancers and the Tarantati, and they were in general very rough; for the sufferers were beaten and goaded in various parts of the body with stones, hammers, swords, clubs, &c., of which treatment the defenders of this extraordinary sect relate the most astonishing examples in proof that severe pain is imperatively demanded by nature in this disorder as an effectual counter-irritant. The Secourists used wooden clubs in the same manner as paviors use their mallets, and it is stated that some Convulsionnaires have borne daily from six to eight thousand blows thus inflicted without danger. One Secourist administered to a young woman who was suffering under spasm of the stomach the most violent blows on that part, not to mention other similar cases which occurred everywhere in great numbers. Sometimes the patients bounded from the ground, impelled by the convulsions, like fish when out of water; and this was so frequently imitated at a later period that the

> women and girls, when they expected such violent contortions, not wishing to appear indecent, put on gowns make like sacks, closed at the feet. If they received any bruises by falling down they were healed with earth from the grave of the uncanonised saint. They usually, however, showed great agility in this respect, and it is scarcely necessary to remark that the female sex especially was distinguished by all kinds of leaping and almost inconceivable contortions of body. Some spun round on their feet with incredible rapidity, as is related of the dervishes; others ran their heads against walls, or curved their bodies like rope-dancers, so that their heels touched their shoulders.

All this degenerated at length into decided insanity. A certain Convulsionnaire, at Vernon, who had formerly led rather a loose course of life, employed herself in confessing the other sex; in other places women of this sect were seen imposing exercises of penance on priests, during which these were compelled to kneel before them. Others played with children's rattles, or drew about small carts, and gave to these childish acts symbolical significations. One Convulsionnaire even made believe to shave her chin, and gave religious instruction at the same time, in order to imitate Paris, the worker of miracles, who, during this operation, and whilst at table, was in the habit of preaching. Some had a board placed across their bodies, upon which a whole row of men stood; and as, in this unnatural state of mind, a kind of pleasure is derived from excruciating pain, some too were seen who caused their bosoms to be pinched with tongs, while others, with gowns closed at the feet, stood upon their heads, and remained in that position longer than would have been possible had they been in health. Pinault, the advocate, who belonged to this sect, barked like a dog some hours every day, and even this found imitation among the believers.

The insanity of the Convulsionnaires lasted without interruption until the year

1790, and during these fifty-nine years called forth more lamentable phenomena that the enlightened spirits of the eighteenth century would be willing to allow. The grossest immorality found in the secret meetings of the believers a sure sanctuary, and in their bewildering devotional exercises a convenient cloak. It was of no avail that, in the year 1762, the Grand Secours was forbidden by act of parliament; for thenceforth this work was carried on in secrecy, and with greater zeal than ever; it was in vain, too, that some physicians, and among the rest the austere, pious Hecquet, and after him Lorry, attributed the conduct of the Convulsionnaires to natural causes. Men of distinction among the upper classes, as, for instance, Montgeron the deputy, and Lambert an ecclesiastic (obt. 1813), stood forth as the defenders of this sect; and the numerous writings which were exchanged on the subject served, by the importance which they thus attached to it, to give it stability. The revolution finally shook the structure of this pernicious mysticism. It was not, however, destroyed; for even during the period of the greatest excitement the secret meetings were still kept up; prophetic books, by Convulsionnaires of various denominations, have appeared even in the most recent times, and only a few years ago (in 1828) this once celebrated sect still existed, although without the convulsions and the extraordinarily rude aid of the brethren of the faith, which, amidst the boasted pre-eminence of French intellectual advancement, remind us most forcibly of the dark ages of the St. John's dancers.

6) Similar fanatical sects exhibit among all nations of ancient and modern times the same phenomena. An overstrained bigotry is in itself, and considered in a medical point of view, a destructive irritation of the senses, which draws men away from the efficiency of mental freedom, and peculiarly favours the most injurious emotions. Sensual ebullitions, with strong convulsions of the nerves, appear sooner or later, and insanity, suicidal disgust of life, and incurable

> nervous disorders, are but too frequently the consequences of a perverse, and, indeed, hypocritical zeal, which has ever prevailed, as well in the assemblies of the Maenades and Corybantes of antiquity as under the semblance of religion among the Christians and Mahomedans.

There are some denominations of English Methodists which surpass, if possible, the French Convulsionnaires; and we may here mention in particular the Jumpers, among whom it is still more difficult than in the example given above to draw the line between religious ecstasy and a perfect disorder of the nerves; sympathy, however, operates perhaps more perniciously on them than on other fanatical assemblies. The sect of Jumpers was founded in the year 1760, in the county of Cornwall, by two fanatics, who were, even at that time, able to collect together a considerable party. Their general doctrine is that of the Methodists, and claims our consideration here only in so far as it enjoins them during their devotional exercises to fall into convulsions, which they are able to effect in the strangest manner imaginable. By the use of certain unmeaning words they work themselves up into a state of religious frenzy, in which they seem to have scarcely any control over their senses. They then begin to jump with strange gestures, repeating this exercise with all their might until they are exhausted, so that it not unfrequently happens that women who, like the Maenades, practise these religious exercises, are carried away from the midst of them in a state of syncope, whilst the remaining members of the congregations, for miles together, on their way home, terrify those whom they meet by the sight of such demoniacal ravings. There are never more than a few ecstatics, who, by their example, excite the rest to jump, and these are followed by the greatest part of the meeting, so that these assemblages of the Jumpers resemble for hours together the wildest orgies, rather than congregations met for Christian edification.

In the United States of North America communities of Methodists have existed for the last sixty years. The reports of credible witnesses of their assemblages for divine service in the open air (camp meetings), to which many thousands flock from great distances, surpass, indeed, all belief; for not only do they there repeat all the insane acts of the French Convulsionnaires and of the English Jumpers, but the disorder of their minds and of their nerves attains at these meetings a still greater height. Women have been seen to miscarry whilst suffering under the state of ecstasy and violent spasms into which they are thrown, and others have publicly stripped themselves and jumped into the rivers. They have swooned away by hundreds, worn out with ravings and fits; and of the Barkers, who appeared among the Convulsionnaires only here and there, in single cases of complete aberration of intellect, whole bands are seen running on all fours, and growling as if they wished to indicate, even by their outward form, the shocking degradation of their human nature. At these camp-meetings the children are witnesses of this mad infatuation, and as their weak nerves are with the greatest facility affected by sympathy, they, together with their parents, fall into violent fits, though they know nothing of their import, and many of them retain for life some severe nervous disorder which, having arisen from fright and excessive excitement, will not afterwards yield to any medical treatment.

But enough of these extravagances, which even in our now days embitter the live of so many thousands, and exhibit to the world in the nineteenth century the same terrific form of mental disturbance as the St. Vitus's dance once did to the benighted nations of the Middle Ages.

www.ReadHowYouWant.com

You can buy our **Large Type** and **EasyRead** books from our www.ReadHowYouWant.com website, from websites like Amazon.com and through your UK and North American bookshop.

EasyRead books are designed to make your reading easy and enjoyable. **EasyRead** books are published in different font sizes, so you can select the font size best for you.

EasyRead is for people with normal eyesight who want books in an easy-to-read format.

EasyRead Comfort is for people who find reading small print tiring but do not need large print.

EasyRead Large is for people who find it easier to read larger print.

The **EasyRead** font, character, word and line spacing have all been set to make it as easy as possible both to recognize a word, and to run your eye along a line of text without losing your place. We split as few words as possible at line ends, as split words make reading harder and can be annoying. For out-of-copyright books, words have been changed to modern spelling (where the sense is not affected), and the original hyphenation of compound words has been retained where this improves word recognition or sense.

With most things you buy, you get a choice, and you can choose the make and model that suits best. There is a big exception – books. You don't get to choose a format that is easy for you to read – you have to read the format the publisher selects.

This **"One Size Fits All"** paradigm **no longer** need apply to books.

At www.ReadHowYouWant.com, you can order a book just as you want it, so you can read it easily. You can choose nearly any format, and at a surprisingly low cost, because of a technical breakthrough that allows us to typeset an individual book automatically, print and bind the book and have it quickly sent directly to you.

You can have your book in **Talking Book** formats (**DAISY or MP3**) or as an **E-book** or in **Braille**.

We have developed totally new visual formats which we hope will help people with **dyslexia** and other print reading disabilities. Look at www.ReadHowYouWant.com for more information.

Good news for publishers and authors. There is a big market for large type, personalized and accessible format books. We can help publishers and authors find new market opportunities for their books, and selling your book through www.ReadHowYouWant.com is easy – we do the work for you. Contact us on info@ReadHowYouWant.com.

Made in the USA
Monee, IL
04 November 2026

67383562R00154